QUALITY OF COMMUNICATION-BASED SYSTEMS

QUALITY OF COMMUNICATION-BASED SYSTEMS

*Proceedings of an International Workshop
held at the TU Berlin, Germany, September 1994*

Edited by

Günter Hommel
Technische Universität Berlin, Germany

KLUWER ACADEMIC PUBLISHERS
DORDRECHT / BOSTON / LONDON

Library of Congress Cataloging-in-Publication Data

```
Quality of communication-based systems : proceedings of an
  international workshop held at the TU Berlin, Germany, September
  1994 / edited by Günter Hommel.
       p.    cm.
   ISBN 0-7923-3259-8 (hb : acid-free paper)
   1. Abstract data types (Computer science)--Congresses.  2. Petri
  nets--Congresses.  3. Parallel processing (Electronic computers)-
  -Congresses.  4. Electronic data processing--Distributed processing-
  -Congresses.   I. Hommel, Günter.
  QA76.9.A23Q35  1995
  004'.35--dc20                                              94-39301
```

ISBN 0-7923-3259-8

Published by Kluwer Academic Publishers,
P.O. Box 17, 3300 AA Dordrecht, The Netherlands.

Kluwer Academic Publishers incorporates
the publishing programmes of
D. Reidel, Martinus Nijhoff, Dr W. Junk and MTP Press.

Sold and distributed in the U.S.A. and Canada
by Kluwer Academic Publishers,
101 Philip Drive, Norwell, MA 02061, U.S.A.

In all other countries, sold and distributed
by Kluwer Academic Publishers Group,
P.O. Box 322, 3300 AH Dordrecht, The Netherlands.

Printed on acid-free paper

All Rights Reserved
© 1995 Kluwer Academic Publishers
No part of the material protected by this copyright notice may be reproduced or
utilized in any form or by any means, electronic or mechanical,
including photocopying, recording or by any information storage and
retrieval system, without written permission from the copyright owner.

Printed in the Netherlands

Program Committee

Hartmut Ehrig
Günter Hommel (Chair)
Stefan Jähnichen
Klaus-Peter Löhr
Bernd Mahr
Miroslaw Malek
Peter Pepper
Radu Popescu-Zeletin
Herbert Weber
Adam Wolisz

Organizing Committee

Uwe Wolfgang Brandenburg
Günter Hommel
Robert Zijal
Armin Zimmermann

Workshop Secretary

Karin Bartscht
Technische Universität Berlin
Sekretariat FR 2–2
Franklinstr. 28/29
10587 Berlin
Germany

Preface

Since 1990 the German Research Society (Deutsche Forschungsgemeinschaft, DFG) has been funding Graduiertenkollegs at selected universities in the Federal Republic of Germany. TU Berlin is one of the first universities that was able to join that new funding program of DFG. The grant amounts to approximately 1.5 million DM for three years. Our Graduiertenkolleg on communication-based systems has been assigned to the computer science department of TU Berlin in 1990 and started its program in October 1991. Although the Graduiertenkolleg has been assigned to TU Berlin the proposal is a joint effort of professors from all three universities in Berlin. The professors responsible for the program are: Hartmut Ehrig (TU Berlin), Günter Hommel (TU Berlin), Stefan Jähnichen (TU Berlin), Peter Löhr (FU Berlin), Bernd Mahr (TU Berlin), Miroslaw Malek (HU Berlin), Peter Pepper (TU Berlin), Radu Popescu-Zeletin (TU Berlin), Herbert Weber (TU Berlin), and Adam Wolisz (TU Berlin).

The Graduiertenkolleg is a Ph.D. program for highly qualified persons in the field of computer science. Twenty scholarships can be granted to fellows of the Graduiertenkolleg for a maximum period of three years. During this time the fellows take part in a selected educational program and work on their Ph.D. thesis. The following research areas are covered by the Graduiertenkolleg:

– Formal specification and mathematical foundations of distributed systems
– Computer networks and multi-media systems
– Software development and concepts for distributed applications
– Language concepts for distributed systems
– Distributed real-time systems
– Reliability, security and dependability in distributed systems

Concurrent and distributed systems have gained increasing importance in numerous new application areas. Those areas are e.g. computer networks, distributed systems in co-operative environments, process control systems, automated manufacturing systems, multi-media systems, and parallel or distributed computer systems.

The development of such systems is of high economical relevance and is currently performed in a network of co-operating institutions from industries, research institutions, and universities. In Germany those activities are supported by national (DFG, Ministry of Research and Technology) and international (ESPRIT, RACE) funding programs. Scientific problems arising during the development of those systems can mostly not be solved sufficiently in the framework of those programs. The need for highly qualified scientists who are able to responsibly direct such projects is growing and there is a lack of education programs leading to such qualifications.

The university education leading to a diploma in computer science gives not enough room for gaining the appropriate knowledge. Only after having acquired a profound knowledge of the classical areas of computer science a specialised education program is suitable. The Graduiertenkolleg allows to fill this gap with its research-oriented scientific education program. The required know-how on technological developments is such completed by scientific questions treated in a selected research and educational program.

For research in the field of communication-based systems the area of Berlin offers best opportunities. The technological infrastructure, research, and development in this field are far advanced compared to the international standard due to numerous ongoing projects in industries, research institutions and universities (e.g. BERKOM and TUBKOM).

The goal of the research program carried out in the Graduiertenkolleg is to study the essential practical and theoretical problems of design, development, and performance and dependability evaluation of communication-based systems. All research carried out in the Graduiertenkolleg is connected with additional projects funded from different institutions. In this way it can be guaranteed that relevant problems for research can be identified and that scientific results flow back into practical solutions.

The goal of the educational program carried out in the Graduiertenkolleg is not only to gain special knowledge in a rather limited field defined by the research area of a fellow but also to gain a broader knowledge on communication-based systems. This should enable the fellows to independently direct research and development projects in that field after having finished their dissertation.

As the goal of the Graduiertenkolleg is ambitious, the admission is highly competitive. Fellows are selected considering the grades of their diploma examination, the duration of their university education and their age, at least one letter of recommendation from a scientific expert, and an interview. During their time in the Graduiertenkolleg a continuing evaluation of their work is performed. Besides the successful participation in the educational program they have to give a report on their scientific progress all six months. After two years the evaluation of the required first draft of the Ph.D. thesis decides on the continuation of the scholarship for the last year. After three years the Ph.D. thesis has to be finished.

This workshop on Quality of Communication-Based Systems is intended to present research results achieved during the first phase of the Graduiertenkolleg to an international community. To stimulate the scientific discussion renown experts have been invited to give their view on the covered research areas. My thanks go to Kurt Jensen, Ugo Montanari, Kishor S. Trivedi, Don Towsley, Nicholas Carriero, and Peter Dickman who accepted our invitation and contributed to the success of this workshop.

Due to the excellent results achieved during the first phase of the Graduiertenkolleg, our external reviewers recommended an extension of the program for another three years starting in October 1994. DFG followed this recommendation and we are looking forward to a next period of hopefully successful research work.

Berlin, July 1994.
Günter Hommel.

Table of Contents

Contributions to Quality from Specification Techniques

Abstract Datatype Semantics for Algebraic High-Level Nets Using
Dynamic Abstract Datatypes 1
Julia Padberg, Technische Universität Berlin

Making Statics Dynamic: Towards an Axiomatization for
Dynamic ADTs 19
Alfonso Pierantonio, Technische Universität Berlin

Specification of Concurrent Systems: from Petri Nets to
Graph Grammars 35
A. Corradini and Ugo Montanari, University of Pisa

Towards a Theory of Strong Bisimulation for the Service Rendezvous 53
Michael Baldamus, Technische Universität Berlin

Contributions to Quality from Quantitative Modelling

Transient Analysis of Real-Time Systems Using Deterministic and
Stochastic Petri Nets 69
Varsha Mainkar and Kishor S. Trivedi, Duke University, Durham

Performance Modeling with Structured Actions 85
Ina Schieferdecker, Gesellschaft Mathematik und
Datenverarbeitung FOKUS

Transient Analysis of Deterministic and Stochastic Petri Nets
by the Method of Supplementary Variables 105
Reinhard German, Technische Universität Berlin

Discrete Time Deterministic and Stochastic Petri Nets 123
Robert Zijal, Technische Universität Berlin

Contributions to Quality from Distributed Systems Organization

Bauhaus Linda: An Overview 137
Nicholas Carriero, David Gelernter and Lenore Zuck,
Yale University, New Haven

Naming and Typing in Languages for Coordination in Open
Distributed Systems 147
Robert Tolksdorf, Technische Universität Berlin

An Efficient Implementation of Decoupled Communication
in Distributed Environments 163
Andreas Polze, Freie Universität Berlin

Extending the Rôle of Object References in Distributed Systems 179
Peter Dickman, University of Glasgow

On Protocols for Loss-less Statistical Multiplexing in
Integrated Networks 181
Mihai Mateescu, Gesellschaft für Mathematik und
Datenverarbeitung FOKUS

Abstract Datatype Semantics for Algebraic High-Level Nets Using Dynamic Abstract Datatypes

Julia Padberg

TU Berlin

Abstract. Using the recently introduced extension of algebraic specifications, called dynamic abstract data types, that is based on algebras as states and transformations of algebras as state transformations, we introduce a new kind of semantics for algebraic high-level nets that especially reflects the data type part. Algebraic high-level nets are interpreted in this framework, where the underlying data type remains unchanged and the net and the net behaviour are transferred into algebras that represent states and dynamic operations relating these algebras. Thus it is possible to give the notions correctness and model semantics in the same way as for algebraic specifications. Nevertheless the abstract data type semantics of algebraic high-level nets is compatible to the usual case graph semantics known from Petri nets in general.

1 Introduction

Petri nets (e. g. [Rei85, Sta90]) are the most common approach to model concurrent and distributed systems within the "true -concurrency" paradigm. This behaviour is given in several kinds of marking graphs, where the nodes represent markings and the arcs the switching of enabled transitions. Since the beginning of the eighties High-Level Petri nets are under investigation. High-Level Petri nets allow structuring either by using sets of colours (Coloured Nets [Jen81]), by logic formulas (Predicate/Transition Nets [GL81]), the programming language ML (Coloured Nets [Jen91]), or abstract datatypes (Algebraic High-Level Nets [Vau86, Hum89]). These variants allow the clear distinction between dynamic and static aspects. In algebraic high-level nets (AHL-nets) the dymanic behaviour is modelled by the net structure, the static aspects by the abstract data type, i. e. an algebraic specification and an algebra.
AHL-nets can be flattened to Place/Transition nets, that behave in the same way as the given AHL-nets, but the complexity of the tokens is expressed by the net structure. By flattening the structure of the algebra is lost, it is merely viewed as a set, which determines the number of places. The semantics of AHL-nets is defined as the case graph of the flattened net. This case graph represents all reachable markings and the switching behaviour of the net. From the Petri net point of view this is sufficient as the case graph allows the usual Petri net analysis. To overcome the loss of the data type, a semantics of AHL-nets is formulated in the sense of algebraic specifications in this paper. The requirements

of such an abstract data type (ADT) semantics for AHL-nets are roughly: distinction of static and dynamic aspects, definition of correctness and verification, compatibility with horizontal and vertical structuring, adequate modelling of concurrency, and distribution and compatibility with Petri net semantics and operational behaviour.

The ADT-semantics is based on a recently developed approach to dynamic representation of algebraic specifications, called dynamic abstract data types (DADT) in [EO94]. This extension of algebraic specifications shall overcome their static concept. Algebraic specifications have been developed since the midseventies, and are one of the main and widely accepted formal specifications. But they have been developed for the description of sequential systems and since a few years there is a distinct interest to extend the static concepts. Among these approaches are D-oids [AZ93, AZ94], Algebraic Specifications with Implicit States [DM94] and DADT's [EO94]. The main idea is to extend the algebraic specifications by some notion of states and state transitions. In the frame of DADT's this is achieved by introducing three levels for syntax as well as for semantics. The first level gives a specification and an algebra as the static part. The next level introduces states by an extended specification and its loose semantics. The third level is given by dynamic operations that relate different states, i. e. the algebras of the second level. The formalism of DADT's is not a fully investigated theory, but a proposed approach, which is still under development. One very promising feature of DADT's is that, although they are an extension of algebraic specifications, other formalisms can and shall be incorporated. This yields a basis to compare and even translate different approaches as various kinds of graph grammars, Petri nets, and others.

Using DADT's as an ADT-semantics we identify the first level with the specification and algebra given in the AHL-net. The second level corresponds to the markings of the net, with the main difference, that the underlying data type is preserved. On the third level the switching of transitions is represented by dynamic operations. Thus at all levels the underlying algebra is given explicitly and with its full structure. Furthermore we present the notion of initial and loose semantics for AHL-net specifications and semantics for AHL-nets. Based upon this it is possible to define correctness w.r.t. some model for Ahl-nets and AHL-net specifications. For the specification of distributed and communication based systems the important impact is that the notion of correctness makes the verification of the specification possible. The verification of specifications surely improves the quality of communication based systems.

In section 2 we review AHL-nets and dicuss the requirements for an ADT-semantics of AHL-nets in detail. Next we give a short introduction of DADT's. In section 4 the construction of the ADT-semantics and related notions and properties are given. In section 5 we discuss, in how far our approach meets the requirements stated in section 2.

2 Requirements for an Abstract Data Type Semantics for Algebraic High-Level Nets

In this section we shortly review AHL-nets and operational behaviour of AHL-nets. Then we discuss the requirements for an ADT-semantics.

2.1 Algebraic High-Level Nets

Based on Place/Transiton nets AHL-nets consist of a net struture and a sprecification together with an algebra. The net is given algebraic as introduced in [MM90]. In this context it is convenient to give the initial marking as a linear sum over ground terms and places. Other formulations are possible and have been used already, but they only differ slightly.

Definition 1 AHL-Net. An algebraic high-level net ,
$N = (SPEC, P, T, pre, post, A, cond)$ consists of a specification $SPEC = (S, OP, E)$, sets P and T (places and transitions, respectively), functions pre, $post$

$$T \xrightarrow[post]{pre} (T_{OP}(X) \times P)^\oplus$$

assigning to each $t \in T$ an element of the free commutative monoid over the cartesian product of terms $T_{OP}(X)$ with variables in X and the set P of places, a $SPEC$-algebra A and a function $cond : T \to \mathcal{P}_{fin}(EQNS(SIG))$ assigning to each $t \in T$ a finite set $cond(t)$ of equations over $SIG = (S, OP)$, the signature of $SPEC$.
The pre ($post$) function assigns to each transition a sum of terms together with their places that are consumed (created) by switching this transition. The set of equations assigned to each transition by $cond$ represents the conditions that must be satisfied for each transition to switch.
The initial marking is given by $m_{init} \in (T_{SPEC} \times P)^\oplus$

Remark. In the context of dynamic abstract data types we will also talk about AHL-net specifications $NSPEC = (SPEC, P, T, pre, post, cond)$ that are AHL-nets without a fixed algebra. Thus the distinction between syntax and semantics is more like in the context of algebraic specifications.

Next we define the operational behaviour of AHL-nets, where we have markings and consistent transitions, where the switching conditions are satisfied.

Definition 2 Switching of Transitions in AHL-nets. Given an AHL-net $N = (SPEC, P, T, pre, post, A, cond)$ we define:

1. The set of place vectors PV, also called marking monoid of N, is the free commutative monoid
$$PV = (A \times P)^\oplus$$
   In this context A is considered to be the disjoint union of all carrier sets of the algebra A, i.e. $A = \biguplus_{s \in S} A_s$.

An element of PV is called a marking of the AHL-net AN.
Especially we have $EVAL_A(m_{init}) \in PV$ where $EVAL_A = (T_{SPEC} \times P)^* \to (A \times P)^\oplus = PV$ is defined on generators by $EVAL_A(term, p) = (eval_A(term), p)$ for each $p \in P$, and $term \in T_{SPEC}$.

2. The set of consistent transition assignments is
$CT = \{(t, asg_A) | t \in T, asg_A : Var(t) \to A\}$ s.t.
A satisfies the equations $cond(t)$ with variables $Var(t)$ under the assginment asg_A
$Var(t)$ is the set of variables that occur in the condition equations $cond(t)$ and in the pre and $post$ conditions $pre(t)$ and $post(t)$ for each transition $t \in T$.

3. The A-induced functions $pre_A, post_A : CT \to PV$ of the AHL-net AN are defined for all $(t, asg_A) \in CT$ by:
$pre_A(t, asg_A) = ASG_A(pre(t))$ and $post_A(t, asg_A) = ASG_A(post(t))$
where $ASG_A = (T_{OP}(Var(t)) \times P)^* \to (A \times P)^* = PV$ is defined on generators by $ASG_A(term, p) = (asg_A(term), p)$ for each $p \in P$, and $term \in T_{OP}(Var(t))$.

4. Given a marking $m \in PV$ and a consistent transition assignment $(t, asg_A) \in CT$ the successor marking $m' \in PV$ of m is defined if $m \leq pre_A(t, asg_A)$ and is computed by

$$m' = m - pre_A(t, asg_A) + post_A(t, asg_A)$$

where $\leq, +, -$ are the usual extensions from natural numbers to linear sums.

Remark. The operational behaviour of AHL-nets can easily be extended to define case graphs with the nodes coloured by markings and the edges by consistent transition assignments.

Example 1 An Airport Information System. As a small example, which is continued in section 4 we present an extract of an AHL-net specification of an airport information system (APIS).
Informal description The data used for the APIS is distributed in order to allow local changes. The data is given mainly by a list of fligths ($\underline{FLIGTHSHEDULE}$), a list of planes ($\underline{PLANESCHEDULE}$) and a list of already booked seats ($\underline{BOOKING}$). Passengers can get information about flights, departure time, etc. Furthermore they are allowed to book fligths by themselves. One main feature is the typical reader/writer problem, concurrent access (lets say 15 at most) is only aloud, if the data is not changed.
The Datat Type The data type is based on an algebraic specification, modelling an aiprport schedule given in [EM90]. The algebraic specification $\underline{APIS}$ is build from several smaller specifications. $\underline{APIS} = \underline{FLIGTHSHEDULE} + \underline{PLANESCHEDULE} + \underline{BOOKING} + ...$ We illustrate parts of $\underline{BOOKING}$.

BOOKING = FLIGTHSHEDULE
        sorts: entry, bookinglist
        opns: create-bk : → bookinglist
                add-bk : entry, bookinglist → bookinglist
                book : flightno bookinglist → bookinglist
                vacant:flightno bookinglist → bool
                ...
        eqns: ...

The Net Structure We only illustrate the nets structure of a subnet BOOKING that models the booking of a flight.

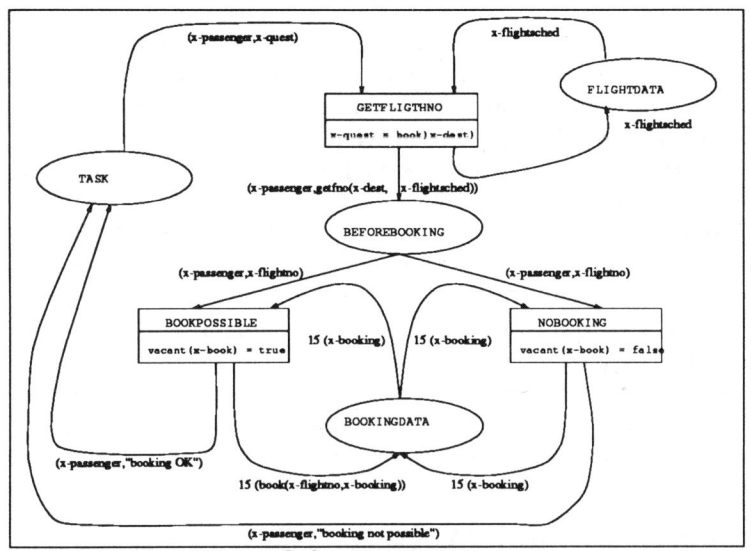

Subnet BOOKING

As the subnet has no initial marking itself, we give some arbitrary one:
$m_{init} =$
$1\S(add(1024, Bonn), add(1025, Rio), create - fs)$, FLIGHTDATA$+$
$1\S(add(1024, true), add(1025, false), create - bk)$, BOOKINGDATA$+$
$((P1, book(Bonn)),$ TASK$) + ((P2, book(Rio)),$ TASK$)$
We have a list of flights (flights to Bonn and Rio), the booking list and two passengers, one wants to go to Bonn, the other to Rio.

2.2 Requirements for an Abstract Data Type Semantics for Algbraic High-Level Nets

The operational behaviour leads to a flattening of AHL-nets into Place/ Transition nets that consist of places PV and transitions CT (see def. 2). Based on this, the semantics of AHL-nets is defined by case graphs. Thus the algebra is given implicitly. From the Petri net point of view this is sufficient but not from the ADT point of view. In ADT's and algebraic specifications behaviour and semantics are based on specific algebras or classes of algebras. But the semantics of AHL-net should not abstract from the dynamic aspects, like processes, concurrency, synchronisation, and distribution. Subsequently some of the main issues for an ADT-semantics for AHL-nets are stated.

1. **Static and Dynamic Aspects** : The ADT-semantics of AHL-nets should adequately reflect and distinguish static and dynamic aspects. The static aspects include classical ADT's and functions which are considered as ADT-operations. The dynamic aspects include transformations between ADT's and processes which are considered as dynamic operations changing the interacting states of the system.
2. **Correctness** The notion of correctness for AHL-nets corresponding to the ADT-semantics with functions and processes should allow to show the correctness of an AHL-net w.r.t. some model. Functions and processes should be given by explicit set theoretic models or other formal specifications.
3. **Horizontal and Vertical Stucturing** : The notions of horizontal and vertical structuring as defined for AHL-nets [PER93] should be compatible with the ADT-semantics of AHL-nets.
4. **Concurrency and Distribution** : The ADT-semantics of AHL-nets should allow to model typical aspects of concurrent and distributed systems, like distributed and shared memory, concurrent access of functions, synchronisation of processes, iterations of processes, and functions as well as deadlock, liveness, and fairness problems.
5. **Net Semantics and Operational Behaviour** : The ADT-semantics of AHL-nets should be compatible with notions for behaviour and semantics of AHL-nets defined using Place/Transition nets.

3 Review of Dynamic Abstract Data Types

In this section we merely recall the concept of DADT's given in [EO94] in order to keep this paper selfcontained.

A <u>dynamic abstract data type</u>, short DADT, consists of an abstract data type, short ADT, together with a collection of <u>dynamic operations</u> which define transformations between the instances of the ADT.

This short proposal allows several different interpretations to different specification techniques. In the following we will be more specific by defining an extended algebraic specification and semantics of DADT's on four different levels, where the first two can be considered to be static and the last two as the dynamic levels.

2.2 4-Level proposal for Dynamic Abstract Data Types

Specification and semantics of a <u>dynamic abstract data type</u>, short DADT, in an extended algebraic version should be given by the following four levels.

Level 1 of DADT's: Value Type Specification

The value type specification VSPEC of a DADT is an algebraic specification with fixed semantics, i. e. the semantics of VSPEC is a monomorphic ADT defined by a given VSPEC-algebra VALUES, e. g. the quotient term algebra T_{VSPEC} defining the initial semantics of VSPEC. Sorts and operations in VSPEC are called value sorts and value operations respectively.

Level 2 of DADT's: Instant Structure Specification

The instant structure specification ISPEC of a DADT is an algebraic specification with constraints in the sense of [EM90] which is an extension of the value type specification VSPEC. The additional sorts and operations in ISPEC are called class sorts and access or attribute functions. The semantics SEM(ISPEC) of ISPEC consists of all ISPEC-algebras A which in general are required to protect the value type, i. e. the restriction af A to VSPEC is isomorphic to VALUES. The algebra A ∈ SEM(ISPEC) are called instant structures and they are intended to model all possible states of the system given by the DADT. For each instant structure A and class sort s the elements in A_s are called objects of class s and denote the existing objects in the state of the system modelled by the instant structure A.

Access and attributed functions [...] are intended to denote the interconnection of objects and the assignment of values to objects in a given state. [...]

Level 3 of DADT's: Dynamic Operations Specification

Dynamic operations of a DADT D are intended to define transformations between instant structures of D, i. e. state transformations. The dynamic operation specification DSPEC of a DADT is not restricted to be an algebraic specification, but may be a suitable extension of algebraic specifications or an integration with other specification techniques [...]. DSPEC includes dynamic operations d_{op}:w and dynamic generators g_{op}:w, where w is a string w= s1,...,sn of sorts s1,...,sn (n ≥ 0) in the instant structure specification ISPEC, i. e. either value sorts or class sorts.

For a given DADT D with instant structure A a dynamic operation d_{op}:w in DSPEC defines for some arguments $a_i \in A_{si}$ (i=1,...,n) a transformation from A to another instant struture B,

$$d_{opD}:(a1,...,an):A \to B$$

For other arguments $a'_i \in A_{si}$ d_{opD}:(a'1,...,a'n) may be undefined or lead to an instant structure B'. If d_{opD}:(a1,...,an) is defined for a given A we in general require that there is at most one resulting structure B (deterministic case). [...]

A dynamic generator g_{op}:w of a DADT D generates for suitable arguments a1,...,an an instant structure A with $a_i \in A_{si}$ (i=1,...,n)

$$g_{opD}:(a1,...,an): \to A$$

[...]

According to the interpretation of dynamic operations d_{op} in a DADT D as discussed above, the basis of the semantics of the specification DSPEC is an <u>instant structure transition system</u> [...].This is a graph, where the set of nodes is the set (or class in the sense of set theory) of all instant structures, i. e. SEM(ISPEC), and the edges are all transformations defined by dynamic operations. If ISPEC is considered to have behavioural semantics we should have a behavioural transition system defined by a quotient of the instant structure transition system due to behavioural equivalence. In general it seems to be useful to consider a <u>transition category</u>, where the objects are suitable equivalence classes of instant structures and the morphisms are transformations generated by dynamic operations. Concurrency analysis of dynamic operations should be possible by analysis of the corresponding transition system or category.

[EO94]

In [EO94] four levels are introduced. But for the purpose of an ADT-semantics of AHL-nets the forth level concerning families of DADT's is - at least at the moment- not relevant.

4 The Abstract Data Type Semantics of Algebraic High-Level Nets

In this section the construction of an ADT-semantics of AHL-nets is presented, leading to notions of initial and loose semantics and correctness. First we summarize the three levels of the DADT's corresponding to an AHL-net. In 4.1, 4.2 and 4.3 we introduce these levels in more detail. Given an AHL-net specification $NSPEC = (SPEC, P, T, pre, post, cond)$ and an initial marking $m_{init} \in T_{SPEC} \times P^\oplus$ the three levels of the DADT are given roughly by

1. **Value Type Specification :** $VPEC = SPEC$ is given by the algebraic specification $SPEC$ of the AHL- net specification $NSPEC$.
2. **Instant Structure Specification :** is an extension of $VSPEC$ w.r.t. some value algebra A. There is new family of sorts $S \times P$ that denotes the class-sorts and an attribute function that includes each class-sort sp into the value sort s.
   The semantics IS is given by algebras $IS \in \textbf{CAT(ISPEC)}$ with $IS_{/VSPEC} = A$ and an amount function $amount : IS_{sp} \to \mathbb{N}$ that yields the number of data elements in one instant structure.
3. **Dynamic Specification :** $DSPEC$ w.r.t. some $A \in \textbf{CAT(ISPEC)}$ consists of a dynamic generator $g : \lambda$ that generates the initial state according to the initial marking m_{init} and dynamic operators $d_{t1...tn} : w$, that are defined on the basis of the switching of one or more transitions in parallel.

4.1 Level 1: Value Type Specification

For an AHL-net specification $NSPEC = (SPEC, P, T, pre, post, cond)$ then the value type specification is $VPEC = SPEC$. The semantics is the usual

ADT-semantics. The quotient term algebra T_{VSPEC} is the initial semantics, and **CAT(VSPEC)** the loose semantics.

4.2 Level 2: Instant Structure Specification

In order to define the amount function and to obtain homomorphisms later we extend the instant specifications by a specification of positive natural number $PNAT$ and a new family of sorts on which the amount function is defined.

ISPEC = VSPEC + PNAT
class-sorts: $S \times P, (S \times P)'$
attributes: $attr_{sp} : sp \to s$
$same_{(sp)'} : (sp)' \to sp$
$amount_{(sp)'} : (sp)' \to nat$

The semantics $SEM_A(ISPEC)$ is given by the class of instant structures IS that respect the value algebra and the natural numbers and where the attributes are inclusions and the families of carriers are in a bijection.

Definition 3. $SEM_A(ISPEC) = \{IS \in \mathbf{CAT(VSPEC)}\}$
with $IS_{/VSPEC} \cong A$ and $S_{/PNAT} = \mathbb{N}$
and $attr_{sp} : IS_{sp} \to IS_s$ is inclusion and $same_{(sp)'} : IS_{(sp)'} \to IS_{SP}$ is bijection.

The definition of instant homomorphisms takes into account that the number of data elements is not a homomorphic property. Thus the homomorphism is on the carrier sets fo $(S \times P)'$ totally undefined.

Definition 4 Instant Homomorphisms. 1. A homomorphism $f : IS_1 \to IS_2$ is a partial homomorphism $f \in Mor^{partial}_{\mathbf{CAT(ISPEC)}}$ with
$f_s = id_{A_s}$ for $s \in S$ the value sorts   and
$attr_{B\ sp} \circ f_{sp} = attr_{A\ sp}$ for $sp \in S \times P$ and
$f_{(sp)'}$ is totally undefined for $(sp)' \in (S \times P)'$
2. Composition is usual composition of partial homomorphisms.

This leads to the category of instant semantics.

Fact 5 Category of Instant Semantics w.r.t. some Algebra.
The category $\mathbf{SEM_A(ISPEC)}$ *consists of the object class* $SEM_A(ISPEC)$ *and instant homomorphisms in between.*

Next (fact 6 - corollary 9) we examine the relation between the semantics of the static part and this level.

Fact 6 Instant Functors. *For each homomorphism*
$h \in Mor_{\mathbf{CAT(VSPEC)}}(A, B)$ *there is a functor* $H : SEM_A(ISPEC) \to SEM_B(ISPEC)$.

Proof. For each $h \in Mor_{\mathbf{CAT(VSPEC)}}(A, B)$ the functor $H : SEM_A(ISPEC) \to SEM_B(ISPEC)$ is defined by :
1. $H(IS_A) = IS_B$ with $IS_{B\ /VSPEC} = B$ and $IS_{B\ /PNAT} = \mathbb{N}^+$ and
$IS_{B\ /S \times P} = \{h \circ attr_{sp}(a) | a \in IS_{A\ /S \times P}\}$

$$IS_{B\ /(S\times P)'} = \{h \circ attr_{sp} \circ same_{(sp)'}(a) | a \in IS_{A\ /(S\times P)'}\}$$
$$amount_{B\ (sp)'} : IS_{B\ (sp)'} \to \mathbb{N}^+ \text{ with}$$
$$amount_{B\ (sp)'}(h(attr(same(a))) = \sum amount_{A\ (sp)'}(a)$$

2. $H(f : IS1_A \to IS2_A) = H(f) : H(IS1_A) \to H(IS2_A)$ with
$H(f)_{sp} : H(IS1_A)_{sp} \to H(IS2_A)_{sp}$ with
$H(f)_{sp}(h \circ attr_{A\ sp}(a)) = h \circ attr_{A\ sp} \circ f(a))$

H is well-defined by definition of $H(IS1_A)_{sp}$ and by compatibility of instant homomorphisms with the attribute function (see def. 4). □

Definition 7 Category of Instant Semantics. The category **SEM(ISPEC)** consists of the class $\{\mathbf{SEM_A(ISPEC)} | A \in \mathbf{CAT(VSPEC)}\}$ as objects and the functors $H : \mathbf{SEM_A(ISPEC)} \to \mathbf{SEM_B(ISPEC)}$ for each homomorphism $h \in Mor_{\mathbf{CAT(VSPEC)}}(A, B)$.

Fact 8. SEM(ISPEC) $\cong$ **CAT(VSPEC)**

Remark. The structure of the categories are isomorphic, NOT the objects. The isomorphism means that this step of the construction of an ADT-semantics preserves the semantics of the static part.

Proof. Follows directly from the definition of **SEM(ISPEC)**

Corollary 9. $\mathbf{SEM_{T_{VSPEC}}(ISPEC)}$ *is initial in* **SEM(ISPEC)**.

Next we give the relation between instant structures and markings. Both represent the state of a system, thus we obtain a bijection between them. The essential difference is that $IS \in \mathbf{SEM_A(ISPEC)}$ is astill structured, i. e. operations and carrier sets ar still existing and furthermore the static and dynamic part are distinguishable. On the other hand $m \in (A \times P)^\oplus$ is merely some word over $A \times P$.

Fact 10. *There is a bijection* $b : (A \times P)^\oplus \to \mathbf{SEM_A(ISPEC)}$.

Proof. Let $b : (A \times P)^\oplus \to \mathbf{SEM_A(ISPEC)}$ be defined in the following way:
for each $pv = \sum_{i=1}^{n} \lambda_i(a_i, p_i) \in (A \times P)^\oplus$ we have
$b(pv) = IS$ with $a \in IS_{sp}$ and $amount_{sp}(a) = \lambda$
$\Leftrightarrow \exists_{j \in 1,\ldots,n} \lambda = \lambda_j$ and $a = a_j$ and $p = p_j$ and $a \in A_s$.
Let $\hat{b} : \mathbf{SEM_A(ISPEC)} \to (A \times P)^\oplus$ be defined by :
$\hat{b}(IS) = \sum_{a \in IS_{sp}} amount(a)(a, p)$
Then obviously $b \circ \hat{b} = id_{\mathbf{SEM_A(ISPEC)}}$ and $\hat{b} \circ b = id_{(A \times P)^\oplus}$ □

4.3 Level 3: Dynamic Operation Specification

Given an AHL-net specification $NSPEC = (SPEC, P, T, pre, post, cond)$ together with an initial marking $m_{init} \in (T_{VSPEC} \times P)^\oplus$, the dynamic specification $DSPEC$ is given by

1. one dynamic generator $g_{init} : \lambda$ yielding the initial state corresponding to the initial marking $m_{init} \in (T_{VSPEC} \times P)^{\oplus}$. This generator has no arguments and yields:
$$g_{init} :\longrightarrow b(EVAL_A(m_{init})) = IS_{init}$$
This is the instant structure that is in bijection to the evaluation of m_{init} in the given algebra.

2. dynamic operations that are defined w.r.t. the parallel switching of transitions. For each $tv \in T^+ = \{t1..tn | n \geq 1 \; ; \; t_i \in T\}$ there is the dynamic operation:
$$d_{tv} : w$$
with $w = w_1...w_n$ and $w_i = sort(x_{i_1})...sort(x_{i_m})$ and $x_{i_j} \in Var(t_i)$. The arguments of a dynamic operation $d_{tv} : a_{1_1}....a_{n_m}$ define assignments $(asg_A^{d_{tv}})_i$ for each t_i ith $(asg_A^{d_{tv}})_i(x_j) = a_{i_j}$.

$$d_{tv} : a_{1_1}....a_{n_m} : Is_1 \to IS_2$$

is defined if:
1. $\sum_{i=1}^n pre_A(t_i, (asg_A^{d_{tv}})_i) \leq \hat{b}(IS_1)$
2. for all $i \in \{1, ..., n\}$ we have $(\overline{asg}_A^{d_{tv}})_i(cond(t_i)) \dashv A$

that means if there are enough data elements (1.) and the switching conditions are satisfied for each transition (2.).
The resulting instant structure IS_2 is computed by:

$$IS_2 = b(\hat{b}(IS_1) - \sum_{i=1}^{n} pre_A(t_i, (asg_A^{d_{tv}})_i) + \sum_{i=1}^{n} post_A(t_i, (asg_A^{d_{tv}})_i)$$

Remark. A special case is $d_t : w$, which corresponds to the switching of one transition.

The instant structure transition system consists of a set of reachable instant structures RIS and the relation given by the dynamic operations TS.

Definition 11 Instant Structure Transition System. Given an AHL-net $NSPEC = (SPEC, P, T, pre, post, A, cond)$ with an initial marking $m_{init} \in (T_{VSPEC} \times P)^{\oplus}$ and let $g_{init} :\to IS_{init}$ and $DYNOP = \{d_{tv} : a1_1...a_{n_m} | tv \in T^+\}$ the instant structure transition system $ISTS_A = (RIS, TS)$ w.r.t. the algebra A is defined by:

1. The least set RIS satisfying
   (a) $IS_{init} \in RIS$ and
   (b) $IS_2 \in RIS$ if $IS_1 \in RIS$ and $d_{tv} : a_{1_1}....a_{n_m} : Is_1 \to IS_2$
2. $TS \subseteq RIS \times DYNOP \times RIS$ with
   $d_{tv} : a_{1_1}....a_{n_m} : IS_1 \to IS_2 \Leftrightarrow (IS_1, d_{tv} : a_{1_1}....a_{n_m}, IS_2) \in TS$

Remark. The recursive definition of the instant structure transition system $ISTS_A$ is analogously to the case graph construction in Petri nets. Together with the bijection $b : (A \times P)^{\oplus} \to \mathbf{SEM_A(ISPEC)}$ (see fact 10) this gives the compatibility of the ADT-semantics with the usual Petri-net semantics.

Fact 12. *For each dynamic operation d_{tv} there is an associated instant homomorphisms $f_{d_{tv}}$.*

Proof. For each $d_{tv} : a_{1_1}....a_{n_m} : IS_1 \to IS_2$ there is $f_{d_{tv}:a_{1_1}....a_{n_m}} : IS_1 \to IS_2$ with
$f_{d_{tv}\ sp}(a) = a$ if $a \in b(\hat{b}(IS_1) - \sum_{i=1}^n pre_A(t_i, (asg_A^{d_{tv}})_i))$
$f_{d_{tv}\ s} = id_{A_s}$
$f_{d_{tv}\ (sp)'}$ is totally undefined.
Obviously $f_{d_{tv}}$ is an instant homomorphism. □

Next we prove that sequential composition of instant homomorphisms associated with dynamic operations is compatible with dynamic operations of parallel transitions. We restrict this fact to the case of two transitions. The extension to $n \in \mathbb{N}$ is straightforward.

Fact 13. *Given some $d_{t_1 t_2} : a_{1_1}....a_{2_m} : IS_1 \to IS_2$, then there is some IS' s.t.*
$d_{t_1} : a_{1_1}....a_{1_m} : IS_1 \to IS'$ *associated with $f_{d_{t_1}}$ and*
$d_{t_2} : a_{2_1}....a_{2_m} : IS' \to IS_2$ *associated with $f_{d_{t_2}}$*
and there is some IS'' s.t.
$d_{t_2} : a_{2_1}....a_{2_m} : IS_1 \to IS''$ *associated with $f'_{d_{t_1}}$ and*
$d_{t_1} : a_{1_1}....a_{1_m} : IS'' \to IS_2$ *associated with $f'_{d_{t_2}}$*
s. t. the subsequent equations hold:
$f_{d_{t_1 t_2}} = f_{d_{t_2}} \circ f_{d_{t_1}}$ *and* $f_{d_{t_1 t_2}} = f'_{d_{t_1}} \circ f'_{d_{t_2}}$

Proof. Given $d_{t_1 t_2} : a_{1_1}....a_{2_m} : IS_1 \to IS_2$,, then there are $(asg_A^{d_{tv}})_1 : Var(t_1) \to A$ and $(asg_A^{d_{tv}})_2 : Var(t_2) \to A$ and t_1 is enabled under $(asg_A^{d_{tv}})_1$. Thus $d_{t_1} : a_{1_1}....a_{1_m} : IS_1 \to IS'$ is defined and IS' is computed by
$$IS' = IS_1 - pre_A(t_1, (asg_A^{d_{tv}})_1) + post_A(t_1, (asg_A^{d_{tv}})_1)$$
We have $IS' \geq IS_1 - pre_A(t_1, (asg_A^{d_{tv}})_1) \geq pre_A(t_2, (asg_A^{d_{tv}})_2)$ and $(asg_A^{d_{tv}})_1(cond(t_1)) \dashv A$ since $d_{t_1 t_2}$ is defined.
Thus $d_{t_2} : a_{2_1}....a_{2_m} : IS' \to IS_3$ is defined and computed by
$$IS_3 = IS' - pre_A(t_2, (asg_A^{d_{tv}})_2) + post_A(t_2, (asg_A^{d_{tv}})_2)$$
$$= IS_1 - pre_A(t_1, (asg_A^{d_{tv}})_1) + post_A(t_1, (asg_A^{d_{tv}})_1)$$
$$- pre_A(t_2, (asg_A^{d_{tv}})_2) + post_A(t_2, (asg_A^{d_{tv}})_2) = IS_2$$
Thus we have $f_{d_{t_1 t_2}} = f_{d_{t_2}} \circ f_{d_{t_1}}$.
The proof for $f_{d_{t_1 t_2}} = f'_{d_{t_1}} \circ f'_{d_{t_2}}$ is similar. □

Definition 14 Transition Category w.r.t. some Algebra. Given an AHL-net $N = (SPEC, P, T, pre, post, A, cond)$ the transition category $\mathbf{TC_A}$ consists of objects $IS \in RIS$ (see def. 11) and the morphisms obtained by the closure of $\{f_{d_{tv}} : IS_1 \to IS_2 | (IS_1, d_{tv} : a_{1_1}...a_{n_m}, IS_2) \in TS\}$.

Fact 15. $\mathbf{TC_A}$ *is subcategory of* $\mathbf{SEM_A}(ISPEC)$

Proof. Due to the definition of $\mathbf{TC_A}$.

Fact 16. *Given the functor* $H : \mathbf{SEM_A(ISPEC)} \to \mathbf{SEM_B(ISPEC)}$ *defined w.r.t.* $h \in Mor_{\mathbf{CAT(VSPEC)}}$ *the restriction to* $H : \mathbf{TC_A} \to \mathbf{TC_B}$ *is well-defined.*

Proof. Due to the definition of H and the fact that SPEC-homomorphisms preserve equations, i. e. for some $h \in Mor_{\mathbf{CAT(SPEC)}}$, $L, R \in T_{SPEC}(X)$ and $asg_A(L) = asg_A(R)$, then $h \circ asg_A(L) = h \circ asg_A(R)$. □

Definition 17 Transition Category. The transition category **TC** has as objects transition categories $\mathbf{TC_A}$ with $A \in \mathbf{CAT(VSPEC)}$ (see def 14), and as morphisms instant functors $H : \mathbf{TC_A} \to \mathbf{TC_B}$ w.r.t. $h \in Mor_{\mathbf{CAT(VSPEC)}}$ (see def. 6).

Fact 18. $\mathbf{TC} \cong \mathbf{SEM(ISPEC)} \cong \mathbf{CAT(VSPEC)}$.

Proof. Given $F : \mathbf{SEM(ISPEC)} \to \mathbf{TC}$ with
$F(\mathbf{SEM_A(ISPEC)} \xrightarrow{H} \mathbf{SEM_B(ISPEC)}) = \mathbf{TC_A} \xrightarrow{H} \mathbf{TC_B}$
and $V : \mathbf{TC} \to \mathbf{SEM(ISPEC)}$ with
$V(\mathbf{TC_A} \xrightarrow{H} \mathbf{TC_B}) = \mathbf{SEM_A(ISPEC)} \xrightarrow{H} \mathbf{SEM_B(ISPEC)}$
Thus $F \circ V = ID_{\mathbf{TC}}$ and $V \circ F = ID_{\mathbf{SEM(ISPEC)}}$ □

Remark. The isomorphism means that the ADT-semantics preserves the semantics of the static part.

Corollary 19. $\mathbf{TC}_{T_{VSPEC}}$ *is initial in* **TC**.

4.4 Semantics of AHL-Nets

Now we define semantics and correctness of AHL-nets and AHL-net specifications. These notions correspond to the ones in ADT's. There loose semantics describes a class of algebras, that satisfy the equations of the specification. In the context of AHL-net specifications it is a class of transition categories, which differ in the representation of the static part.

Definition 20 Loose Semantics of AHL-Net Specifications.
Given an AHL-net specification $NSPEC = (SPEC, P, T, pre, post, cond)$ the loose semantics is defined by the transition category **TC**.

The initial semantics in ADT's is the quotient term algebra, that is the generated algebra, which satisfies only the equations given in the specification. Initial semantics of AHL-nets is the transition category that is based on the quotient term algebra of the static part.

Definition 21 Initial Semantics of AHL-Net Specifications. Given an AHL-net specification $NSPEC = (SPEC, P, T, pre, post, cond)$ the initial semantics is defined by $\mathbf{TC}_{T_{VSPEC}}$, the transition category w.r.t. T_{VSPEC}.

As the algebras of a AHL-net is already fixed, we merely obtain the semantics based on this algebra, which belongs to the loose semantics of the AHL-net specification.

Definition 22 Semantics of AHL-Nets. Given an AHL-net $N = (SPEC, P, T, pre, post, A, cond)$ the semantics is defined by $\mathbf{TC_A}$, the transition category w.r.t. A.

The semantics given above are quite restrictiv, as they do not allow equivalences on the dynamic part. A sensible extension is to investigate behavioural equivalences imlied by behavioural specifications.

The notion of semantics lead to correctness. Correctness is defined with respect to some given model. This model can be either given explicitly as a set theoretic model or by some other formalism. Especially the incorporation of other formalisms into DADT's leads to interesting possibilities for the description of models.

Definition 23 Correctness of AHL-Net Specifications. Given an AHL-net specification $NSPEC = (SPEC, P, T, pre, post, cond)$ $NSPEC$ is correct w.r.t. some model MC if $\mathbf{TC_{T_{VSPEC}}} \cong \mathbf{MC}$.

Definition 24 Correctness of AHL-Nets. Given an AHL-net $N = (SPEC, P, T, pre, post, A, cond)$ N is correct w.r.t some model MC if $\mathbf{TC_A} \cong \mathbf{MC}$

Example 2 Semantics of the APIS. We discuss the third level only. The generator $g_{init} :\longrightarrow IS_{init}$ generates the following instant structur IS_{init}:

$IS_{init\ /APIS} = T_{APIS}$
$IS_{init\ (task,TASK)'} =$
$IS_{init\ task,TASK} = \{(P1, book(Bonn)), (P2, book(Rio))\}$
$IS_{init\ (flightsched,FLIGHTDATA)'} = IS_{init\ flightsched,FLIGHTDATA} =$
$\{(add(1024, Bonn), add(1025, Rio), create - fs))\}$
$amount_{(lightsched,FLIGHTDATA)'}((add(..., create - fs))) = 15$
$IS_{init\ (booking,BOOKINGDATA)'} = IS_{init\ booking,BOOKINGDATA} =$
$\{(add(1024, true), add(1025, false), create - bk))\}$
$amount_{(booking,BOOKINGDATA)'}((add(..., create - bk))) = 15$
All other carrier sets are empty.

Next we give the result of the dynamic operation $d_{\mathbf{GETFLIGHTNO}}$ that represents the switching of **GETFLIGHTNO**.

$d_{\mathbf{GETFLIGHTNO}} : (P1, book(Bonn)), (add(..., create - fs)) : IS_{init} \to IS_1$ with
$IS_1\ _{/APIS} = T_{APIS}$
$IS_1\ _{(task,TASK)'} = IS_{init\ task,TASK} = \{(P2, book(Rio))\}$
$IS_1\ _{(flightsched,FLIGHTDATA)'} = IS_1\ _{flightsched,FLIGHTDATA} =$
$\{(add(1024, Bonn), add(1025, Rio), create - fs))\}$
$amount_{(lightsched,FLIGHTDATA)'}((add(..., create - fs))) = 15$
$IS_1\ _{(booking,BOOKINGDATA)'} = IS_1\ _{booking,BOOKINGDATA} =$
$\{(add(1024, true), add(1025, false), create - bk))\}$

$amount_{(booking,BOOKINGDATA)'}((add(..., create - bk))) = 15$

$IS_{1\ (befbook,\textbf{BEFOREBOOKING})'} = IS_{1\ befbook, \textbf{BEFOREBOOKING}}$
$= \{(P1, 1024)\}$

All other carrier sets are empty.

As the example is small, there is only one process "booking of a flight", that is represented by $d_{\textbf{GETFLIGHTNO}}$, $d_{\textbf{BOOKPOSSIBLE}}$, and $d_{\textbf{NOBOOKING}}$. In this case the process does not have parallel actions, in general this would be expressed by some dynamic operation $d_{t_1...t_n}$ based on parallel transitions.

5 Adequacy of the Abstract Data Type Semantics of Algebraic High-Level Nets

In this section we discuss in how far the introduced ADT-semantics meets the requirements stated in section 2.

1. **Static and Dynamic Aspects :** Distinction between static and dynamic aspects is clearly given by the distinction between the static value type specification $VSPEC$ and the $VSPEC$ algebra on the one hand (see 4.1) and the dynamic aspects given by the instant structure specification $ISPEC$ and the dynamic specification $DSPEC$ on the other (see 4.2 and 4.3)
2. **Correctness :** is defined in subsect. 4.4 corresponding to the usual ADT-semantics.
3. **Horizontal and Vertical Stucturing :** An open problem is the compatibility of structuring techniques known from AHL-nets with the ADT-semantics of AHL-nets. That means that the same notions like union, fusion, and transformation should be defined for the semantics and have to be respected by the construction of the ADT-semantics. To obtain this first morphisms between DADT's have to defined and a category of DADT's must satifay some properties, e. g. cocompleteness.
4. **Concurrency and Distribution** is preserved due to the definition of dynamic operations based on parallel transitions. An open problem is the formulation of typical Petri net notions, especially the techniques to analyse the net and the semantics, have to be transferred into this context. Nevertheless this should not cause severe problems due to the compatibility with the case graph of the flattened net (see fact 10 and remark to def. 11).
   The dynamic operations are based on the parallel switching of transitions, thus conurrency is expressed implicitly. A sensible extension is to define explicit parallel dynamic operations, which are compatible to the implicit parallel dynamic operations, we have defined in this paper.
5. **Compatibility between ADT-Semantics and Case Graph** is given (see fact 10 and remark to def. 11). The operational behaviour extends the term-rewriting of algebraic specifications due to the definition of dynamic operations based on transitions.

6 Conclusion

In this paper we have presented an approach to an ADT-semantics of AHL-nets, that is based on DADT's. This opens a door to the important results and well-known methods known from ADT's. Here we have defined a semantics so that the static and dymanic parts of an AHL-net are reflected in the way that the initial semantics of the net preserves the initial semantics of the algebraic specification given in the AHL-net. This semantics is defined for AHL-net specification in the two folded ADT-style, i. e. with a loose and an initial semantics. AHL-nets already include a fixed algebra, thus the semantics is also fixed. In both cases the ADT-semantics is compatible with the case graph semantics and the operational behaviour of AHL-nets. Thus the typical Petri net aspects are preserved. Based on the semantics we can define a correctness w.r.t. some model. This notion is very important from a practical point of view. Correctness notions render the verification of specifications. Verification is an important way to improve the quality of specifications and thus the quality of the specified system. Moreover the framework of DADT's gives the possibility to describe the models not only in a unstructured set theoretic way, but also using other well-developed specification formalisms.

We have given and discussed some main requirements for such an ADT-semantics, many of them are satisfied. Nevertheless this is the first step to such an ADT-semantics of AHL-nets, thus there is still research to be done. We have pointed out some ideas how to solve the open problems.

References

[AZ94] E. Astesiano and E. Zucca. D-oids: A model for dynamic data types. *Special Issue of MSCS,*, 1994. accepted for pulication.

[AZ93] E. Astesiano and E. Zucca. A semantic model for dynamic systems. *Springer Workshops in Computing*, pages 63–80, 1992/93.

[DM94] P. Dauchy and Gaudel M.G. Algebraic specifications with implicit states. *Tech. Report, Univ. Paris Sud*, 1994.

[EM90] H. Ehrig and B. Mahr. *Fundamentals of Algebraic Specification 2: Module Specifications and Constraints*, volume 21 of *EATCS Monographs on Theoretical Computer Science*. Springer, Berlin, 1990.

[EO94] H. Ehrig and F. Orejas. Dynamic abstract data types: An informal proposal. *Bull. EATCS 53*, 1994.

[GL81] H.J. Genrich and K. Lautenbach. System modelling with high-level Petri nets. *Theoretical Computer Science*, 13:109–136, 1981.

[Hum89] U. Hummert. *Algebraische High-Level Netze*. PhD thesis, 1989.

[Jen81] K. Jensen. Coloured petri nets and the invariant method. *Theoretical Computer Science*, 14:317–336, 1981.

[Jen91] K. Jensen. *Coloured Petri Nets. Basic concepts, analysis methods and practical use*. Springer-Verlag, Berlin, 1991.

[MM90] J. Meseguer and U. Montanari. Petri nets are monoids. *Information and Computation*, 88(2):105–155, oct. 1990.

[PER93] J. Padberg, H. Ehrig, and L. Ribeiro. Algebraic high-level net-transformation systems. Technical Report 93-12, 1993. Revised Verion accepted for Mthematical Structures in Computer Science.

[Rei85] W. Reisig. *Petri nets*. Springer Verlag, 1985.

[Sta90] P. Starke. *Analyse von Petri-Netz-Modellen*. Teubner Verlag, 1990.

[Vau86] J. Vautherin. Parallel systems specification with coloured Petri nets and algebraic abstract data types. In *Proc. of the 7th European Workshop on Application and Theory of Petri nets*, pages 5–23, 1986.

Making Statics Dynamic:
Towards an Axiomatization for Dynamic ADTs

Alfonso Pierantonio*

TFS, Fachbereich Informatik
Technische Universität Berlin
Franklinstr. 28/29 (Sekr. FR6-1)
D-10587 Berlin, Deutschland

Abstract. The paper describes a new mathematical structure able to model the dynamics of complex systems such as object communities. The states of the system is modeled by configuration algebras. Evolutions allow the system to change his current state via state updating and object creation/deletion. The dynamic behavior is induced from the static specification of classes.

1 Introduction

In the last years a number of approaches have been proposed to better understand and formalize those concepts which made object oriented programming successful in the practical side of software design. Among these approaches many exploit fruitfully algebraic and categorical methods (e.g. see [3, 8, 14] just to mention a few). These approaches usually formalize the concepts of class and inheritance using Abstract Data Types specification techniques. This provides all the semantic theory of a very well known framework which has been investigated for about 20 years.

Unfortunately, classical algebraic specification fails in describing more complex but essential key notions such as object *identity*, *dynamics*, and *persistency*. Usually objects are models of real-world concepts: they are referred to as having structure, behavior and identity. The structure and behavior of similar objects are described by a class. The object identity is such a property which distinguishes two objects existing at the same time, e.g. two stacks are different entities, although having the same specification which describes the common properties. Such entities survive at the execution of a method or a function unless they are explicitly asked to die: pushing/popping elements into/from a stack let it evolve from a given state to another but won't destroy it. Objects have therefore a persistent nature.

This work proposes an informal introduction to a mathematical formalism which allows the modeling of object identities and object values in the same setting. A state of the object system is given by an algebra. The object values are determined by the algebraic class specification. While the class extension, i.e. the set of existing objects at a given time (in the sense of object oriented database), depends on the

---
* The research is supported in part by the European Communities under grant HCM ERBCHBICT930300 and by DFG.

evolution of the system. Each object has a state which is encoded in such algebras. The object system can evolve in time because of, for instance, the execution of a method on a given object. Such an evolution results in a modification of the state of the object which received the message, i.e. the object execution is interpreted by a transformation of an algebra into another.

There is also a more general motivation for such a work. In fact, in the last time a number of new approaches towards the specification of systems with a dynamic behavior has been introduced. They try to propose more elegant solutions to the problem of the state representation. The idea of formalizing the notion of states as algebras is presented in [10] although this approach has been more used for the semantics of programming languages. This has lately triggered a number of activities tending towards a more uniform treatment of complex concepts as types and type instances. Among these, it is worthwhile to mention the work based on the notion of *implicit state* by Dauchy and Gaudel [4] and the work based on a identity-bookkeeping mechanism called *tracking map* in the *d-oids* by Astesiano and Zucca [2]. The former relies more on the syntactical level while the latter is essentially without any syntax.

The work presented in this paper is meant to provide a semantical framework in which all these issues are addressed consistently. The originality of the approach consists of deriving the dynamic behavior of the system from the static specification of the classes. This lead to an abstract specification of dynamics in contrast with the *d-oids* where there is no axiomatization at all; in [4] although the state is implicit, the specification of the dynamics is a bit verbose and at a low level with the so-called elementary modifiers. Another important point is the abstract treatment of the identity, i.e. not referring to any identification mechanism which represents the way in which the identity is implemented. In the programming languages area usually variable names are used to keep object distinguished mixing *addressability* and identity. In the data base area object are kept distinguished through attributes which identify a tuple, mixing *data values* and identity [11]. In our approach we address the problem of the identity without mixing it with other notions.

Our work together with the already mentioned approaches (i.e. [2, 4]) and the work of Große-Rhode [9] about the specification of parallel state dependent systems, motivated the recent formulation of Dynamic Abstract Data Types by Ehrig and Orejas [7]. The main difference in such a version of DADT is that the dynamic operations, i.e. those which transform a state into another, are derived from the static part. Such a peculiarity provides a big deal of abstraction in contrast with the fact that dynamic operations are often specified in an imperative manner.

Furthermore, this work is the natural continuation of previous work ([14, 15, 16]). In such works, the concepts of class and inheritance has been deeply investigated and analyzed and compared. Moreover, a formal theory of design reuse has been proposed.

The paper is organized as follows, in the next section some basic notions about algebraic specification are recalled. The section 3 introduces the class model and some of the results which have been already achieved are described. Section 4 is devoted to the object dynamics which is illustrated via a number of examples. In the last section, some conclusions are drawn.

2 Preliminary Notation

In this section, we briefly review some basic notions on the algebraic specifications; details can be found in [5]. A *signature* Σ is a pair (S, OP) where S is a set of *sorts* and OP a set of *constant* and *function symbols*; constant symbols are considered as operation symbols of arity 0. A *pointed signature* is a signature $\Sigma = (S, OP)$ with a distinguished element of the set S of sorts denoted by $pt(\Sigma)$. By a Σ-*algebra* $A = (S_A, OP_A)$ of a signature $\Sigma = (S, OP)$ we mean two families $S_A = (A_s)_{s \in S}$ and $OP_A = (N_A)_{N \in OP}$, where A_s are sets for all $s \in S$, which are called *domains* of A, and $N_A : A_{s_1} \times \cdots \times A_{s_n} \longrightarrow A_s$ are functions for all operator symbol $N : s_1 \cdots s_n \longrightarrow s$ and all $s_1 \cdots s_n \in S^+$, $s \in S$ (for constant symbols $N : \longrightarrow s$, $N_A \in A_s$). The set of all Σ-algebras is denoted by $Alg(\Sigma)$.

If $\Sigma_1 = (S_1, OP_1)$ and $\Sigma_2 = (S_2, OP_2)$ are signatures, a *signature morphism* $h : \Sigma_1 \longrightarrow \Sigma_2$ is a pair of functions $(h^S : S_1 \longrightarrow S_2, h^{OP} : OP_1 \longrightarrow OP_2)$ such that for each $N : s_1 \cdots s_n \longrightarrow s$ in OP_1 and $n \geq 0$ we have $h^{OP}(N) : h^S(s_1) \cdots h^S(s_n) \longrightarrow h^S(s)$ in OP_2. A signature morphism $h : \Sigma_1 \longrightarrow \Sigma_2$ induces a *forgetful functor*

$$V_h : Alg(\Sigma_2) \longrightarrow Alg(\Sigma_1)$$

defined, for each Σ_2-algebra A'', by $V_h(A'') = A' \in Alg(\Sigma_1)$ with $A'_s = A''_{h^S(s)}$ for each $s \in S_1$, $N_{A'} = h^{OP}(N)_{A''}$ for each $N \in OP_1$. A *pointed signature morphism* is a signature morphism $h : \Sigma_1 \longrightarrow \Sigma_2$ such that $h^S(pt(\Sigma_1)) = pt(\Sigma_2)$. It is easy to check that pointed signature morphisms are closed under composition.

By an *algebraic specification* $SPEC = (\Sigma, E)$ we intend a pair consisting of a signature Σ and a set E of (positive conditional) equations. If $SPEC_1 = (\Sigma_1, E_1)$ and $SPEC_2 = (\Sigma_2, E_2)$ are two algebraic specifications, a *specification morphism* $f : SPEC_1 \longrightarrow SPEC_2$ is a signature morphism $f : \Sigma_1 \longrightarrow \Sigma_2$ such that the translation $f^\#(E_1)$ of the equations of $SPEC_1$ is contained in E_2. A *pointed algebraic specification* is an algebraic specification with a pointed signature. A *pointed specification morphism* between pointed specifications is a pointed signature morphism f such that $f^\#(E_1) \subseteq E_2$.

For notational convenience, when $SPEC = (\Sigma, E)$ is a pointed specification the distinguished sort $pt(\Sigma)$ will be also denoted by $pt(SPEC)$.

The algebraic specifications and the specification morphisms form the category CATSPEC of algebraic specifications [5]. The CATSPEC category is closed with respect to pushouts and pullbacks.

3 The Class Model

In this section we propose a class specification in order to model with generality the class notion present in the current object oriented programming languages. Such a mathematical notion could be fruitfully exploited to better understand the main mechanisms and their interactions.

The importance of inheritance is widely recognized, but it is not the only peculiar feature of the object oriented paradigm: encapsulation is considered as important [1]. This protection mechanism allows to trace a boundary between the implementation and the outside. The operations which can be invoked over the instances of a class are just those listed in the external interface of the class itself and any attempt at executing a private operation results in an error.

The minimalization of the interdependencies of separately written software components and the reduction of the amount of implementational details are among the major benefits due to this technique. The concept of abstract data type is strengthened by the presence of an external interface because of the separation of the functionalities versus the implementation of an abstraction.

Unfortunately, inheritance can reduce the benefits of encapsulation. In fact accessing, in a subclass, inherited variables leaves the designer of the superclass unable to rename, reinterpret or remove these variables. On the other hand, there are two categories of clients of a class, those who need to manipulate objects (the clients of the instances) and those who want to reuse somehow the class in order to specialize it or just reuse its code (the clients of the class) [18]. Thus, many languages have two external interfaces, one for each kind of clients, and they are usually defined incrementally since the interface for the class clients has a greater view than the instance clients. We will call these interfaces *instance* and *class interface* (see the table 1).

Usually, inheritance allows to arrange classes in specialization hierarchies through a *top down* process. Moreover, we can also define a class by reusing the code of another one or instances from other classes. This results in a *bottom-up* process. An interesting case arises if we allow an explicit import interface where requiring some features which are necessary in order to realize the external behavior of the class. This allows us to implement a class assuming the existence of some functionalities but disregarding which class will provide them. Unfortunately, most programming languages have constructs for defining modules and reuse them only by listing their names. This approach goes in the opposite direction of abstraction since we need to declare explicitly which is the supplier of a particular abstraction.

The class notion is the combination of two different concepts: the module and the type concepts. The class specification is therefore composed of a certain number of parts: a parameter part, an import interface, an instance and a class export interface. All these components declare signatures and their properties. Moreover, in the external interfaces there is the class sort, a sort of interest which describes the concept the class is intended to model. More formally

Definition 1 Class Specification and Semantics. A class specification C_{spec} consists of five algebraic specifications PAR (parameter part), EXP_i (instance interface), EXP_c (class interface), IMP (import interface) and BOD (implementation part) and five specification morphisms as in the following commutative diagram.

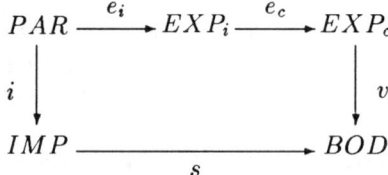

The specification EXP_i, EXP_c and BOD are pointed specifications, and e_c and v are pointed specification morphisms.

The semantics $SEM(C_{spec})$ of a class specification is the set of all pairs (A_I, A_{E_c}) of algebras, where $A_I = V_s(A_B)$ and $A_{E_c} = V_v(A_B)$ for some $A_B \in Alg(BOD)$. The pointed sort $pt(EXP_i)$ is called class sort.

Interpretation Each of the five parts consists not only of signatures, but also of equations, which describe some of the properties of the operations. We distinguish among constructors, which are zero arity functions, methods, which have the form

$$\text{meth:s } s_1 \cdots s_n \longrightarrow s \tag{1}$$

and functions

$$\text{fun:s } s'_1 \cdots s'_n \longrightarrow s' \tag{2}$$

The interfaces EXP_i and EXP_c describe the external access functions and their behavior: the former describes the messages which can be sent to the objects that are instances of the class, while the latter contains the methods which can be used by other classes. The part of BOD not in EXP_c is hidden from other classes. The specification BOD describes an implementation of the exported methods using the ones provided by the IMP specification. The import specification IMP contains information about *what* is needed by BOD to implement EXP_c, but not *which* class can provide it: the latter task is provided by the interconnection mechanisms. The specification PAR models genericity, unconstrained if the specification consists of sorts only, constrained when the parameter is required to have operations satisfying certain properties.

Remark. The semantics chosen for a class specification is only one of a number of possibilities and the theory presented here can be developed by choosing one of the alternatives. One possibility is a functorial semantics (as done in [14]) where the meaning of a class specification is a (functorial) transformation which takes a model (an algebra in this case) of IMP and returns a model (again an algebra) of EXP_c. In the latter case, the semantics must be definable uniformly for all classes and must be closed with respect to Amalgamation ([5]). Another possibility is a loose semantics, where only some pairs (A_I, A_{E_c}) are chosen. In the former case, what are needed are results such as the Extension Lemma in [5] if, for example, the semantics is the composition of the forgetful functor $V_v : Alg(BOD) \longrightarrow Alg(EXP_c)$ with the free functor $Free_s : Alg(IMP) \longrightarrow Alg(BOD)$.

Definition 2 Class. A class $C = (C_{spec}, C_{impl})$ consists of a class specification C_{spec} and a class implementation C_{impl} such that $C_{impl} = A_B$ for some BOD–algebra A_B.

Example 1. The morphisms in this example are just inclusions. In the notation we use the keywords **Parameter, Instance Interface, Class Interface, Import Interface, Body** to declare the subspecification to be added to the parts already defined. For example, since $PAR \subseteq EXP_i$, after the keyword **Instance Interface** only $EXP_i - PAR$ is listed. When a subspecification keyword is missing, the relative specification is just the closest subspecification in the diagram. In the instance interface the distinguished sort $pt(EXP_i)$ is stated by the *class sort* clause.

Next, the STACK class specification is defined. Such an abstract data type is a *stack* which allows to push and pop elements.

Stack is Class Specification
Parameter
   <u>sort</u>   data
   <u>opns</u> $\bot$:⟶data
Instance Interface
   <u>class sort</u> stack
   <u>opns</u>   EMPTY:⟶stack
              PUSH:stack data⟶stack
              POP:stack⟶stack
              TOP:stack⟶data
   <u>eqns</u>   POP(PUSH(s,x)) = s
              TOP(PUSH(s,x)) = x
Class Interface
   <u>sort</u>   nat
   <u>opns</u>   0:⟶nat
              _+1:nat⟶nat
              _.HEAD:stack⟶nat
   <u>eqns</u>   EMPTY.HEAD= 0
              PUSH(s,x)).HEAD= s.HEAD+1
Import Interface
   <u>sorts</u>   array,nat
   <u>opns</u>   0:⟶nat
              _+1:nat⟶nat
              NIL:⟶array
              _[_]:=_:array nat data⟶array
              _[_]:array nat⟶data
   <u>eqns</u>   NIL[i] = $\bot$
              ($a[i]:=e$)[j] = **if** $i=j$ **then** e **else** $a[j]$
              ($a[i]:=e_1$)[i]:=e_2 = $a[i]:=e_2$
Body
   <u>opns</u>   <_,_>:array nat:⟶stack
   <u>eqns</u>   EMPTY= <NIL,0>
              PUSH(<a,n>,x) = <$a[n]:=x,n+1$>
              TOP(<$a,n+1$>) = $a[n]$
              POP(<$a,n+1$>) = <$a[n]:=\bot,n$>
              <a,n>.HEAD= n
End Stack

The STACK example also shows the rôle played by each component of the class specification. The instance interface describes the *abstract* properties, in the sense that the inteded behavior is representation independent. The body describes, in turn, how such data type is implemented in terms of what is required by the formal import. In other word, we suppose the imported items as predefined and then we use them for *programming* the features of the class. The constructor

$$<\_,\_>:\text{array nat}\longrightarrow\text{stack}$$

describes the record consisting of the instance variables, i.e. we intend to realize the stack by means of an array and a natural as a pointer. The equations in the body specify how the value of the *representation record* change consistently with the export specification. Of course, if one would like to represent the stack in a different way, such constructor will be defined accordingly, e.g.

$$< \_ >: list \longrightarrow stack$$

in case the representation is based on a linked list. The syntax use the underscore as a placeholder for the arguments, e.g.

$$\_[\_]:=\_ : array\ nat\ data \longrightarrow array$$

can be used on an array a with parameters n and x in this way $a[n]:=x$. The element $\bot: \longrightarrow data$ is required since we need a unique representation for each stack, i.e. our semantical framework does not allow us to implement the pop operations just decreasing the pointer of the stack. This problem can be overcame relying on our institution independent formalism and adopting a behavioral semantics setting.

Presenting this class model we intend to cover a large number of class structures as they are defined in current object oriented languages. We have focused on the importance of avoiding uncontrolled code reuse without any constraint, and therefore, in the proposed model, we provide an explicit import interface. None of the languages analyzed allows to specify some requirements for the import, although some allow the direct importing of other existing classes, incorporating (with the *use* clause) a combination mechanism. The opportunity to hide some implementational aspects gives to a class designer the freedom to modify the implementation without affecting the clients of the instances of that class. All the languages analyzed but BETA have constructs for protection of data representation. The set of all public operations of a class forms the external interface, which we call *instance interface*. Another form of protection is given to prevent another kind of client, the designer of a subclass, to access some variables. We have named this other interface, which contains the instance one, *class interface*. The C++, POOL, Trellis/Owl have an explicit class interface, distinct from the instance interface. For instance, in the C++ language the instance interface consists of all public items which can be declared like that via a *public* clause, while the class interface includes both the public and the subclass visible items, declared through a *protected* clause, accessible only to derived classes.

Encapsulation and inheritance are the major features of object oriented methodology but other techniques can as well enhance some quality factors. In [12] an informal comparison between genericity and inheritance is presented. Genericity represents a good solution to achieve a good amount of flexibility with a static type system (untyped languages provide a great deal of flexibility, but all the errors can be detected only at run-time). Many languages, such as Eiffel, Trellis/Owl, POOL, BETA, and OOZE, allow genericity, although with some differences. The only properties treated by these languages are signature properties; OBJ allows one to specify behavior with an equational language, by means of theories and views, while OOZE use pre and post conditions in the style of Z. All these languages supply an actualization mechanism in order to instantiate the generic classes.

In table 1 we have indicated which of the components of our model of class are explicitly present in the notion of class in some of the analyzed languages. In this table we have distinguished among unconstrained and constrained genericity (which are indicated by the *constr.* and *unconstr.* keywords). The former kind of genericity does not allow constraint (operations) on the generic type parameters; the latter one does.

	Instance Interface	Class Interface	Genericity	Formal Import	Specialization Inheritance	Reusing Inheritance
BETA	NO	NO	*constr.*	NO[a]	YES	NO
C++	YES	YES	NO[b]	NO	YES	YES
Eiffel	YES	NO[c]	*unconstr.*[d]	NO[a]	YES	NO
POOL	YES	YES	*constr.*	NO[a]	NO	YES
Smalltalk	YES	NO[e]	NO	NO	YES	NO
Trellis/Owl	YES	YES	*constr.*	NO[a]	YES	YES

[a] Although it is possible to import through actualization, there is no explicit import interface.
[b] We refer to the version ANSI 2.0.
[c] The class interface coincides with the implementation part.
[d] We refer to the version described in [12, 13].
[e] The class interface coincides with the instance interface.

Table 1. Language Analysis

It is worthwhile mentioning how inheritance has been modeled in the framework as well (see [14, 15, 16]). We distinguished between *specialization* and *reusing inheritance* relations. The former is a technique for the functional specialization of the external behavior of a class: its use corresponds to a public declaration of the subclass designer that the instances of the subclass obey to the semantics of the superclass. The latter is a technique for the reuse of the code of the superclass: its use corresponds to a private decision of the subclass designer to reuse the code of the superclass without any constraint or type compatibility. As shown in table 1 by means of the model we classified the inheritance concept as it has been implemented in the analized languages.

An important result shows how the specialization inheritance can be used to simulate the reusing inheritance [14, 15]. This is important when inheritance is used as a requirement design technique for completing incomplete specification: non-monotonic design decisions can therefore be reduced to monotonic ones via generalization. In [16] an inheritance operator has been introduced. Such an operator induces the specialization inheritance relation. Moreover, the semantics of the subclass is expressed in terms of the semantics of the superclass and of the semantics of the modifications due to the inheritance. In essence, the subclass reuses not only the code of the superclass but also its properties and verifications. The operators have the Local Confluence and Compatibility properties which are the basis for a theory of design reuse.

4 Object Dynamics

This section is devoted to the dynamic behavior of an object system. Since we are interested in presenting the general idea and the underlying rationale we will illustrate this part by means of an example. More complete and formal definitions and results can be found in [17].

A Class Specification as we already saw describes the *statics* of a class, i.e. all those properties which are common to all the instances and which can be referred to as the declarative semantics of the class. It makes sense to expect that the behavior of the objects depends strictly on the static part. Let us consider a class as defined in the previous section, i.e. a pair consisting of a class specification and an algebra as class implementation. The method symbols are interpreted by the functions in the algebra. These functions manipulate the values which can be assumed by the objects of the class. The values do not represent any object unless we consider single values as very simple and basic objects. In essence, a class specification determines which are the values that the instances of a class can have and how to pass from a value to another by means of the functions.

The basic idea is to extend the semantic model in order to cover also the notion of object. Objects are entities with structure, behavior and identity. Since we want to treat objects with an implicit state in the sense of Dauchy and Gaudel ([4]) only behavior and identity are considered. Dealing with objects which are abstract, i.e. without structure, is widely recognized to be important for reasoning about them. At a given time, each object has a current state value which is an element of the algebra associated with the class specification. More generally, we intend to define an *object transition system* (OTS) in which every state is represented by an algebra. Such algebras, which are called *instant structures* in the terminology of *d-oids* [2], contain

1. the objects existing in such a state,
2. the values the objects can assume,
3. an actual state function which provides for each object its current value.

For each class sort s we need to represent both objects and their values: we distinguish therefore between the sorts s^{Ω} and s^v. Furthermore, a state function associates for each object its current values as depicted in fig. 1. The diagram refers to only one sort but in general the number of sort is arbitrary and a state corresponds to a family of such structures. In such an example, only three objects are existing (obj_1, obj_2 and obj_3), A_s^{Ω} denotes the class extension for the class sort s, A_s^v denotes all the possible values of the class sort s although in the figure only the current values of the object are evidenced (v_1, v_2 and v_3). In each state of the OTS the value of the objects can range over the same set of values. This implies that the instant structures have the same object values part which is fixed. The object values are those which are determined by the class implementation as we already said.

The OTS is formalized with the mathematical notion of category. The objects of such a category are the states of the OTS, i.e. the instant structures, while the morphisms of the category model the evolutions. It is important to mention how such morphisms are not compatible with the state functions in the instant structures. The reason is that a state changing over an instant structure has a *locality property* which

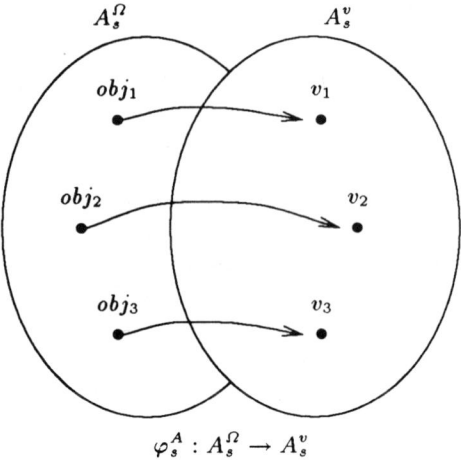

Fig. 1. An object system state

changes the algebraic structure of the instant structure itself. Evolutions from a state to another occur whenever there is a *method execution* or an *object creation/deletion*. For the sake of semplicity, we will consider here only method executions.

Usually the specification is obtained after a design process which results in a class topology as in fig. 2. The classes are related by means of refinement relations,

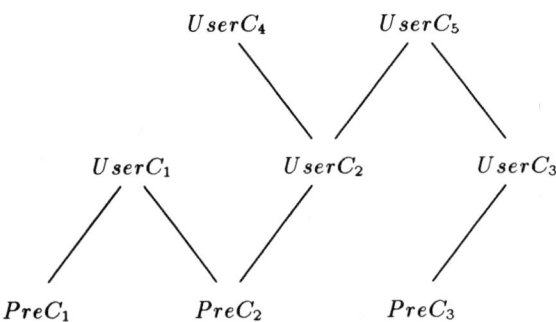

Fig. 2. A Class Topology

e.g. inheritance, or by means of clientship relations, e.g. paramer passing or composition. In the diagram, the classes $PreC_i$ are the predefinite classes, whereas the classes $UserC_i$ are the user–designed classes. From such a design stage we intend to obtain the specification of object values and the algebra of the object values. In the case of algebraic class specification we compute the union of the instance interfaces of the classes present in the diagram, obtaining $SPEC^v$. Then for each class we consider the correspondent implementation algebra and compute the amalgamation sum according to the same diagram, obtaining A^v.

Example 2. Consider the class specification STACK given in the previous section. Replacing the formal parameter by the natural numbers specification we obtain the class specification of *stacks of naturals*. Of such an actualized class specification we consider only the instance interface. Such a specification, say **stack**^{v}, follows

stack^v =
sorts   stack, nat
opns   0:⟶nat
              _ +1:nat⟶nat
              EMPTY:⟶stack
              PUSH:stack nat⟶stack
              POP:stack⟶stack
              TOP:stack⟶nat
eqns   POP(PUSH(s,x)) = s
              TOP(PUSH(s,x)) = x

The specification here listed is the specification of values which is used to illustrate the dynamics of objects of sort stack.

In order to introduce the OTS category we need to show how to extend canonically the object values specification $SPEC^v$ towards a configuration specification $SPEC^\varphi$. Therefore, for each sort s in $SPEC^v$ we obtain two sorts s^v and s^Ω, where s^v is a renaming of s, while s^Ω is the sort of the objects of sort s. In few words, for each sort we want to have besides the object values carrier set (given by A^v) the identities carrier set. $SPEC^\varphi$ will contain also the state function symbols

$$\left(\varphi_s^v : s^\Omega \longrightarrow s^v\right)_{s^v \in S^v} \qquad (3)$$

Formally $SPEC^\varphi$ is obtained as the pushout object of $SPEC^v$ and SIG^φ with respect of S^v according to the following diagram where SIG^φ contains the sorts S^v

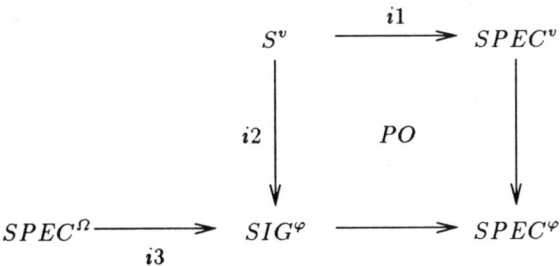

and S^Ω, and state functions (3).

Example 3. Consider the specification **stack^v**, its canonical extention to the configuration specification **stack$^\varphi$** is

stack$^\varphi$ =
sorts   stack^v, stack$^\Omega$, nat^v, nat$^\Omega$
opns $0: \longrightarrow \text{nat}^v$
   $\_+1 : \text{nat}^v \longrightarrow \text{nat}^v$
   $\text{EMPTY}: \longrightarrow \text{stack}^v$
   $\text{PUSH}: \text{stack}^v\ \text{nat}^v \longrightarrow \text{stack}^v$
   $\text{POP}: \text{stack}^v \longrightarrow \text{stack}^v$
   $\text{TOP}: \text{stack}^v \longrightarrow \text{nat}^v$
   $\varphi_{\text{nat}}: \text{nat}^\Omega \longrightarrow \text{nat}^v$
   $\varphi_{\text{stack}}: \text{stack}^\Omega \longrightarrow \text{stack}^v$
eqns $\text{POP}(\text{PUSH}(s,x)) = s$
   $\text{TOP}(\text{PUSH}(s,x)) = x$

The operation symbols $\varphi_{\text{nat}}: \text{nat}^\Omega \longrightarrow \text{nat}^v$ and $\varphi_{\text{stack}}: \text{stack}^\Omega \longrightarrow \text{stack}^v$ will be interpreted by state mappings, i.e. functions which associate each object with its current value, e.g. as the function $\varphi_s^A : A_s^\Omega \longrightarrow A_s^v$ in fig. 1.

The dynamic behavior of the objects is given as a categorical transition system. We are going now to describe how the components of such a category look like.

Given a configuration specification $SPEC^\varphi$ and the object values algebra A^v we define the OTS as the category **Trans**(A^v) which is the model where interpreting the method executions.

Objects. The objects of the category **Trans**(A^v) represents the state of the OTS. The object of the category are quadruples of algebras as follows

$$S = (A3, A2, A0, A1) \tag{4}$$

where

- $A1 = A^v \in Alg(SPEC^v)$ is the object values algebra
- $A0 \in Alg(S^v)$
- $A2 \in Alg(SIG^\varphi)$
- $A3 \in Alg(S^\Omega)$

with

- $V_{i2}(A2) = A0$ with $i2 : S^v \longrightarrow SIG^\varphi$
- $V_{i1}(A1) = A0$ with $i1 : S^v \longrightarrow SPEC^v$
- $V_{i3}(A2) = A3$ with $i3 : S^\Omega \longrightarrow SIG^\varphi$

In other word, for each instant structure we require that the amalgamation sum holds such that we can consider the algebra

$$A1 +_{A0} A2 \in Alg(SPEC^\varphi)$$

as a state of the system at a given time .

Example 4. Consider the configuration specification **stack**$^\varphi$ given in the example 3 and as algebra of values A^v the initial algebra. Please note that the algebra of values is a model of the specification **stack**^{v}, i.e. $A^v \in Alg(\textbf{stack}^v)$. We want to consider now two stacks existing at the same time, say st_1 and st_2 with state values $empty$ and $push(empty, 0)$, respectively. It is important to stress how the elements of A^Ω_{stack}, i.e. st_1 and st_2 are not identifiers: they are identities. The component which corresponds to the SIG^φ specification, denoted by $A2$ in the previous definition, is depicted in fig. 3. In general, a state of the OTS being a model of the specification **stack**$^\varphi$ has

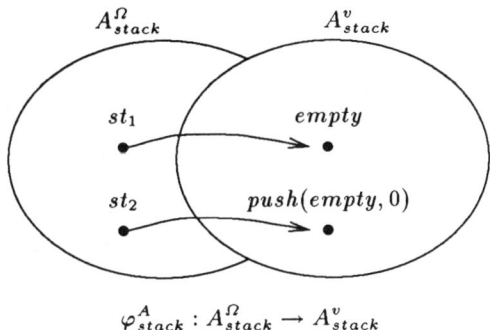

Fig. 3. Two stack objects.

also some structure which is not present in the component $A2$.

Morphisms. The morphisms of the category $\textbf{Trans}(A^v)$ are defined in such a way they can model the object evolutions. Given two instant structures

$$S = (A3, A2, A0, A1) \text{ and } S' = (A3', A2', A0', A1')$$

then there is a morfism $m : S \longrightarrow S'$ which is a quadruple of mappings

$$(\phi|_{S^\Omega} : A3 \longrightarrow A3', \phi : A2 \longrightarrow A2', id_{A0} : A0 \longrightarrow A0, id_{A1} : A1 \longrightarrow A1)$$

and a family of functions

$$(f^o_s : A^v_s \longrightarrow A^v_s)_{o \in A2^\Omega_s}$$

briefly

$$< (\phi|_{S^\Omega}, \phi, id_{A0}, id_{A1}), (f^o_s)_{o \in A2^\Omega} >$$

such that

$$\varphi^{A2'}_s(\phi(o)) = f^o_s(\varphi^{A2}_s(o)) \quad (5)$$

for all identities $o \in A2^\Omega_s$ and for all $s \in S^v$. The amalgamation sum conditions over S' implies that ϕ does not change the values part of $A2$, it can only affect the object carrier sets $A2^\Omega_s$ and the state functions $\phi^{A2}_s : A2^\Omega_s \longrightarrow A2^v_s$.

The category **Trans**(A^v) as it has been defined is parameterized over the object values algebra A^v which can range over $Alg(SPEC^v)$. This means that for each different object values algebra $A \in Alg(SPEC^v)$ a category **Trans**(A) is defined.

The category **Trans**(A^v) can be used as intepretation for the object system evolutions. We distinguish among methods execution, object creation and deletion.

Method Execution. Let $S = (A3, A2, A0, A1) \in$ **Trans**(A^v) be a state of the OTS. Let $\omega \in A2_s^\Omega$ be an existing object of sort s and let $m : ss_1 \cdots s_n \longrightarrow s$ a method symbol in $SPEC^v$. Executing m over ω with parameters $a_1 \in A1_{s_1}^v, \ldots, a_n \in A1_{s_n}^v$ causes an evolution described by the morphism

$$eval : S \longrightarrow S'$$

defined as

$$< (id_{A3}, id_{A2}, id_{A2}, id_{A0}, id_{A1}), (f_s^o)_{o \in A2_s^\Omega} >$$

where

$$f_s^o = \begin{cases} \lambda x.m^{A1}(x, a_1, \ldots, a_n) & \text{if } o = \omega \\ id_{A1_s^v} & \text{otherwise} \end{cases} \quad (6)$$

The identities in *eval* just say that the carrier sets will remain the same, no object is created or deleted. While (6) says that the only object ω will change is current value.

Example 5. Consider the fig. 3 in the example 4. Pushing the natural $s(s(0))$ on the stack object st_1 results in a changing of the state, i.e. in a transition which leads to the configuration in fig. 4. This is modeled by keeping unchanged the class extension, i.e. there are not any created or deleted object, and by the following family of functions

$$f_{stack}^{st_1}(x) = \lambda x.push(x, s(s(0))) \text{ and } f_{stack}^{st_2} = id_{A_{stack}^v}$$

The situation is summarized in fig. 5 where it is shown how the state mapping

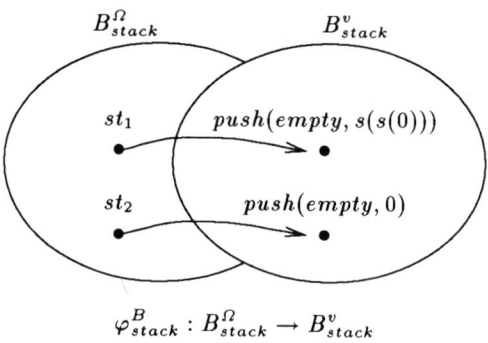

Fig. 4.

changes after the execution of the method accordingly with the condition (5).

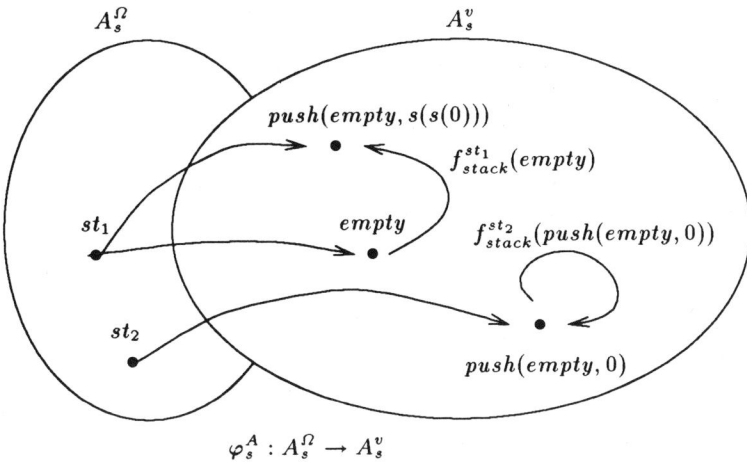

Fig. 5.

5 Conclusions

In this paper an original approach to the formalization of the dynamics of complex system is proposed. The main idea is that type instances are considered to inherit the properties described in their correspondent ADT specifications. This means that the basic specifications are enough to define also the operational behavior of the type instances. This work represents a specific version of Dynamic Abstract Data Types in the sense of Ehrig and Orejas ([7]). In general, the specification of dynamic operations of an DADT, i.e. operations which model state changings, requires an extra level (the third one). The specification of such dynamic operations is usually pretty imperative (see e.g. [9]) while we keep on having an abstract view on them. Our approach allows to deal with objects which have an abstract state: they consists of identity, behavior but not structure. But at the same time, the semantical framework allows one to keep track of the side-effects due to the execution of methods.

Other advantages of this work is that we can have a time varying dimension of the state while in the work of Dauchy and Gaudel [4] the state is fixed. With respect to the *d-oids* we present the possibility of axiomatize the dynamic behavior while the d-oids have only a semantical nature.

Acknowledgments. I am grateful to Francesco Parisi-Presicce for all the valuable discussions and his support. I wish also to thank Uwe Wolter for his interest in my work. Last but not least, Philipp W. Kutter is gratefully acknowledged for the excellent drawings.

References

1. America,P.: *Designing an Object-Oriented Programming Language with Behavioral Subtyping.* Proc. REX/FOO L, Springer LNCS 489, 1990, pp. 60-90

2. Astesiano,E., Zucca,E.: *D-oids: A Model for Dynamic Data Types.* Special Issue of MSCS, to appear 1994
3. Breu,R.: *Algebraic Specification Techniques in Object Oriented Programming Environments.* Springer LNCS 561, 1991
4. Dauchy,P., Gaudel,M.C.: *Algebraic Specification with Implicit State.* Techn. Report, Univ. Paris–Sud, 1994
5. Ehrig,H., Mahr,B.: *Fundamentals of Algebraic Specification 1. Equations and Initial Semantics.* EATCS Monograph in Computer Science, Vol.6, Springer Verlag, 1985
6. Ehrig,H., Mahr,B.: *Fundamentals of Algebraic Specification 2. Module Specifications and Constraints.* EATCS Monograph in Computer Science, Vol.21, Springer Verlag, 1990
7. Ehrig,H., Orejas,F.: *Dynamic Abstract Data Types: An Informal Proposal.* Bull. EATCS 53, June 1994
8. Goguen,J., Diaconescu,D.: *Towards an Algebraic Semantics for the Object Paradigm.* Proc. 10th WADT, Springer LNCS, 1994
9. Große–Rhode,M.: *Specification of Parallel State Dependent Systems.* TU Berlin, 1994, in preparation
10. Gurevitch,Y.: *Evolving Algebras, A Tutorial Introduction.* Bull. EATCS 43, 1991, pp.264–284
11. Khoshafian, S., Copeland, G.: *Object Identity.* Proc. OOPSLA 1986, ACM Press, 1986, pp. 406–416
12. Meyer,B.: *Genericity versus Inheritance.* Proc. OOPSLA 1986, ACM Press, 1986, pp. 391–405
13. Meyer,B.: *Object-Oriented Software Construction.* Prentice-Hall, 1988
14. Parisi–Presicce,F., Pierantonio, A.: *An Algebraic Approach to Inheritance and Subtyping.* Proc. ESEC 1991, Springer LNCS 550, 1991, pp. 364–379
15. Parisi–Presicce,F., Pierantonio, A.: *An Algebraic Theory of Class Specification.* ACM Transaction on Software Engineering and Methodology, accepted for publication 1994
16. Parisi–Presicce,F., Pierantonio, A.: *Reusing Object Oriented Design: An Algebraic Approach.* Proc. International Symposium on Object–Oriented Methodologies and Systems, to appear 1994
17. Parisi–Presicce,F., Pierantonio, A.: *Dynamical Behavior of Object Systems.* In preparation
18. Snyder,A.: *Encapsulation and Inheritance in Object–Oriented Programming Languages.* Proc. OOPSLA 1986, ACM Press, 1986, pp. 38–45
19. Weber,H., Ehrig,H.: *Specification of Concurrently Executable Modules and Distributed Modular Systems.* Proc. Workshop Future Trends of Distr. Comp. Systems in the 1990's, Hong Kong, 1988, pp. 202–215
20. Wegner,P.: *Dimensions of Object–Based Language Design.* Proc. OOPSLA 1987, ACM Press, 1987, pp. 168–182. Also as special issue of SIGPLAN No. 22, 12, Dec 1987

Specification of Concurrent Systems: from Petri Nets to Graph Grammars*

A. Corradini and U. Montanari

Università di Pisa, Dipartimento di Informatica, Corso Italia 40, 56125 Pisa, Italy
({andrea,ugo}@di.unipi.it)

Abstract. We first review some aspects of Place/Transition Petri nets, which are the basis of their success as a specification formalism for concurrent and distributed systems. In particular, we summarize some results concerning the truly-concurrent semantics of safe nets, stressing the fruitful use of categorical techniques.
Next we discuss the use of Graph Grammars (according to the algebraic, double-pushout approach) as a specification formalism, showing that they are strictly more expressive than P/T nets. We also describe the state of the art of research activities aimed at providing graph grammars with a categorical truly-concurrent semantics.

1 Introduction

The nets which owe their name to Carl Adam Petri [Pet62, Rei85] have been the first formal tool proposed for the specification of the behaviour of systems which are naturally endowed with a notion of concurrency. The success of Petri nets in the last thirty years can be measured by the looking not only at the uncountably many practical applications of nets, but also at the developments of the theoretical aspects, which range from a complete analysis of the various phenomena arising in simple models of nets to the definition of more expressive (and complex) classes of nets.

Such a success of Petri nets as specification formalism for concurrent or distributed systems is due (among other things) to the fact that they can describe in a natural way the evolution of systems whose states have a distributed nature. In fact, thinking for example to the so-called Place/Transition nets, a state of the system to be specified is represented by a *marking*, i.e., a set of *tokens* distributed among a set of *places*. Thus the state is intrinsically distributed, and this makes easy the explicit representation of phenomena like *mutual exclusion*, *concurrency*, *sequential composition* and *non-determinism*.

While for sequential, deterministic systems an input/output semantics is often satisfactory, for concurrent or reactive systems (which are intrinsically non-deterministic) such a semantics is usually not sufficient. Indeed, in general one desires a more complete description of the relationships among the elementary

---

* Research partially supported by the COMPUGRAPH Basic Research Esprit Working Group n. 7183

steps of a computation, possibly including information about causality, concurrency, or about the points where non-deterministic choices were taken. Such semantics may for example help the understanding of the operational behaviour of a net, can be used to analyze the relationships with other nets or with other systems, or can play an important role in building bigger systems from the composition of elementary ones. As a matter of fact, Petri nets have been equipped along the years with rich, formal computation-based semantics, including both interleaving and truly-concurrent models. In most cases such semantics were defined via well-established categorical techniques, often involving adjunctions between suitable categories.

After introducing the basics of Place/Transition Petri nets, in Section 2.1 we will present informally various possible computation-based semantics for a sample (one-)safe net [NPW81], including the construction of the corresponding *prime event structure*: such structures are widely accepted as a good semantic domain for systems exhibiting concurrency and non-determinism. In Section 2.2 we will summarize the categorical techniques and results presented in [Win87], where the prime event structure semantics of safe nets is presented in a very elegant way as a chain of adjunctions among suitable categories. We will take the opportunity to stress the relevance of the use of categorical techniques in semantics, and the advantages of regarding systems "in-the-large", i.e., of considering the semantics not of an isolated system, but of a whole *category* of them.

After this informal summary of some basic aspects of Petri nets, we will introduce *graph grammars* (or *graph rewriting*) as a useful generalization of nets. We will stick to the algebraic, double-pushout approach to graph grammars [Ehr87]. It is not difficult to explain in which sense we consider graph rewriting as strictly more powerful than P/T nets. After all, although the state of a net has an intrinsically distributed nature, its structure in undoubtedly poor. Formally, it is just a multiset of places, or, equivalently an isomorphism class of sets (of tokens) labelled by places. Thus a net is just a "(multi)set rewrite system". Even if this is sufficient for many purposes, it is easy to believe that in more complex situations the state of a distributed system cannot be described faithfully just as a *set* of components, because also *relationships* among components should be represented. Thus graphs turn out to be more natural for representing distributed states. As a consequence, graph rewriting appears as a suitable formalism for describing the evolution of a wide class of systems, whose states have a natural distributed and interconnected nature. Therefore it is of great interest on the one hand to study in depth the relationship between graph rewriting and Petri nets, and on the other hand to equip graph grammars with semantics which generalize the corresponding semantics for nets.

However, in the literature graph grammars have been considered almost exclusively as a generalization of string grammars or of term rewriting systems to the rewriting of more complex structures. As a consequence, although the algebraic theory of graph grammars comprises many results concerning parallelism and concurrency (see [Kre77, Kre87, Ehr87]), most of those results recast in this more general framework notions and results of (term) rewriting systems, exploring properties like confluence, Church-Rosser, orthogonality of redexes, parallel moves, and so on. Only recently some truly-concurrent semantics of graph grammars (based on prime event structures) have been proposed [Sch94, CELMR94b], but much more has to be done to arrive to a theory comparable with that of Petri nets. In particular, it is still missing an "in-the-large" categorical approach, able to provide semantics of grammars via suitable adjoint functors: in fact, only recently the problem of defining a *category of graph grammars* has been considered [CELMP94a, CELMP94b].

In the second part of the paper, after introducing some basic notions related to graph grammars in Section 3.1, we make precise in which sense they are a generalization of P/T Petri nets in Section 3.2. Finally in Section 4 we describe the "state of the art" in the semantics of concurrency for algebraic graph grammars, and hint at some expected results.

2 Place/Transition Petri Nets

As stated in the introduction, Petri nets are widely accepted as an adequate formalism for the specification of concurrent/distributed systems. Indeed, the *state* of a net, i.e., a set of *tokens* distributed among a set of *places*, has an intrinsically distributed nature. As a consequence, nets can specify in a natural way phenomena like *mutual exclusion, concurrency, sequential composition* and *non-determinism*. Moreover, they have a pleasant graphical presentation, which makes their use appealing also for the non technical user. In this paper we focus on the class of Place/Transition Petri nets, and on its subclass of safe nets, borrowing the basic definitions from [Rei85].

Definition 1 (static aspects of Place/Transition Petri nets). A (infinite capacity) **marked Place/Transition Petri net** (shortly **(P/T) net**) is a tuple $\mathbf{N} = \langle S, T, F, W, \underline{M} \rangle$ where

- S is a set of **places**;
- T is a set of **transitions** (we require that $S \cap T = \emptyset$);
- $F \subseteq (S \times T) \cup (T \times S)$ is a relation, called the **causal dependency relation**;
- $W : F \to \mathbb{N}^+$ is a **weight** function; we consider W as extended canonically to $(S \times T) \cup (T \times S)$ by $W(x,y) = 0$ iff $(x,y) \notin F$; and
- $\underline{M} : S \to \mathbb{N}$ is an **initial marking** function. □

In the sample net of Figure 1 the places are $S = \{A, B, C, D, E\}$ (drawn as circles), the transitions are $T = \{a, b, c, d, e\}$ (represented as thick line segments), the causal dependency relation F is represented by arrows (for example, $(A, a), (c, D) \in F$, but $(d, D) \notin F$). The weight function is $W(x, y) = 1$ for all $\langle x, y \rangle \in F$ (thus weights are not explicitly depicted). The initial marking is $\underline{M}(A) = 1$ and $\underline{M}(X) = 0$ for $X \neq A$. In general, a marking $M : S \to \mathbb{N}$ is represented pictorially by a set of $M(X)$ *tokens* (black dots) in each place $X \in S$.

The operational behaviour of a net is described by the so-called "token game". A transition is enabled to fire if there are enough tokens in its "preconditions": in

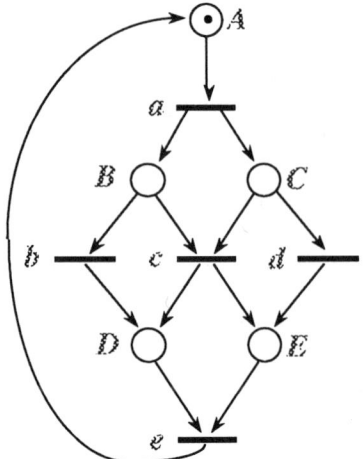

Fig. 1. A safe Place/Transition Petri net

this case its firing removes some tokens from the preconditions and creates some new tokens in its "postconditions", according to the weight function. Moreover, more transitions can fire simultaneously.

Definition 2 (dynamic aspects of P/T nets [Rei85]). Let $\mathbf{N} = \langle S, T, F, W, \underline{M} \rangle$ be a P/T net. Then

- A function $M : S \to \mathbb{N}$ is called a **marking** for $\mathbf{N}$.
- For each transition $t \in T$, its **precondition** $pre(t)$ is the marking defined as $pre(t)(s) = W(s,t)$, and its **postcondition** $post(t)$ is the marking defined as $post(t)(s) = W(t,s)$.
- A transition $t \in T$ is **enabled** at M iff[2] $pre(t) \leq M$.
- If $t \in T$ is a transition which is enabled at marking M then t may **occur** (or **fire**), yielding a new marking M' defined as $M' = M - pre(t) + post(t)$. This situation is denoted by $M[t\rangle M'$.
- The pre and $post$ functions can be extended in the obvious way to multisets of transitions in T. Then the definitions of the enabling conditions and of the effect of firing apply unchanged to multisets of transitions. □

In the net of Figure 1 transition a is the unique enabled in the initial marking. Its firing deletes the token in A, and generates a new marking, say M_1, having one token in B and one in C. In marking M_1 there are three enabled transitions: b, c, and d. Moreover the (multi)set $\{b, d\}$ is enabled as well, but neither $\{b, c\}$ nor $\{c, d\}$ are (they would need two tokens in B and C, respectively). Thus b and d can be said to be *concurrently enabled*, while for example b and c are *in*

---

[2] If M and M' are markings, $M \leq M'$, $M + M'$, and $M - M'$ are meant to be computed pointwise. Thus $M \leq M' \Leftrightarrow (\forall s \in S \,.\, M(s) \leq M'(s))$; $(M + M')(s) = M(s) + M'(s)$ for all $s \in S$, and if $M' \leq M$, then $(M - M')(s) = M(s) - M'(s)$ for all $s \in S$.

conflict or *mutually exclusive*: the firing of one of the two prevents the firing of the other. After the firing of either $\{b, d\}$ or c we obtain marking M_2 having one token in D and on in E. Transition e is enabled in M_2, and its firing produces the initial marking. Thus the net has a cyclic behaviour.

The net of Figure 1 is of a particular kind. It is called **safe** because every marking reachable from the initial one has at most one token in each place (thus it is actually a *set*, not a proper multiset). Most of what follows will be presented for safe nets only, because the treatment is easier, but can be generalized to arbitrary P/T nets.

2.1 Computation-based Semantics

For sequential systems it is often sufficient to consider an input/output semantics (thus often the semantic domain consists of a suitable class of functions). For concurrent/distributed systems, in the semantics one often wants to record more information about the actual computation(s) performed by the system: e.g., one may want to know which steps of a computations are independent (concurrent), or which are causally related. For example, such information is necessary if one wants to compose concurrent systems, keeping the semantics compositional, or if one wants to allocate a computation on a distributed system.

There are many computation-based semantics for safe nets. They differ for the amount of information one wants to record in the semantics, and for the way it is recorded. Let us assume that one desires to record in a single structure (some abstraction of) all the possible computations of a net starting from the initial marking. Then there are two orthogonal dimensions along which the various semantic domains differ: non-determinism and concurrency.

Non-determinism. Nets are intrinsically nondeterministic devices, because of the mutual exclusion phenomenon. Qualitatively, there are two main ways for giving semantics to a nondeterministic system:

Set based: Collect all the possible computations of a system in a *set*, forgetting about the choice point during computations.

Branching structure. Collect all the possible computations of a system in a *branching structure* (e.g., a tree), which also records at which points of the computations certain choices have been made.

Concurrency. Concurrency can be considered as a primitive, directly observable notion, or one may assume that observers are sequential, and thus the concurrent happening of two events only means that in two "linearizations" of a computation the two events can appear in any order. Thus the following two approaches are possible:

True concurrency. The fact that two events are "independent" or "not causally related" is represented directly in the semantic domain using a partially ordered structure.

Interleaving. Each computation of the system is represented by a set (or by a tree, depending on the representation of non-determinism) of sequences of events. The independence of two events in a computation must be represented by two different sequences where the two events appear in different orders. Thus concurrency is reduced to non-determinism.

In Table 1 we indicated some possible semantic domains for nets, for the four possible combinations of the above parameters.

Non-determinism \ Concurrency	**Interleaving**	**True concurrency**
Set-based	(Set of) Firing Sequences	(Set of) Deterministic processes
Branching structure	Tree of firing sequences	Non-deterministic processes
		Event structures

Table 1. Some possible semantic domains for safe nets

For the safe net of Figure 1, the set of firing sequences is easily generated: $\{a, ab, ad, ac, abd, adb, ace, \ldots\}$. More compactly, this set can be described by the regular expression $(a(bd + db + c)e)^*$. Figure 2 shows (part of) the tree of firing sequences of the same net, which is obtained by gluing together the common prefixes of the firing sequences.

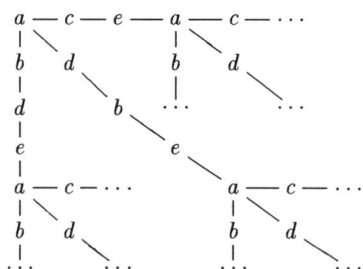

Fig. 2. The tree of firing sequences of the net of Figure 1

In the left side of Figure 3 we show one deterministic non-sequential process of the net of Figure 1. A deterministic process of a net N is an acyclic net (also called *Occurrence Net*) without forward and backward conflicts (i.e., each place has at most one ingoing and one outgoing arc), together with a net homomorphism to the original net; in Figure 3 the net homomorphisms are indicated by labeling places and transitions of the occurrence net with places and transitions

of the net of Figure 1. A deterministic process enjoys the property that if we put
one token on every "minimal" place (i.e., in the running example on the topmost place labelled by A), then in the resulting marked net every firing sequence
individuates uniquely a firing sequence of the original net (through the net homomorphism), and the causal dependencies among transitions are preserved.

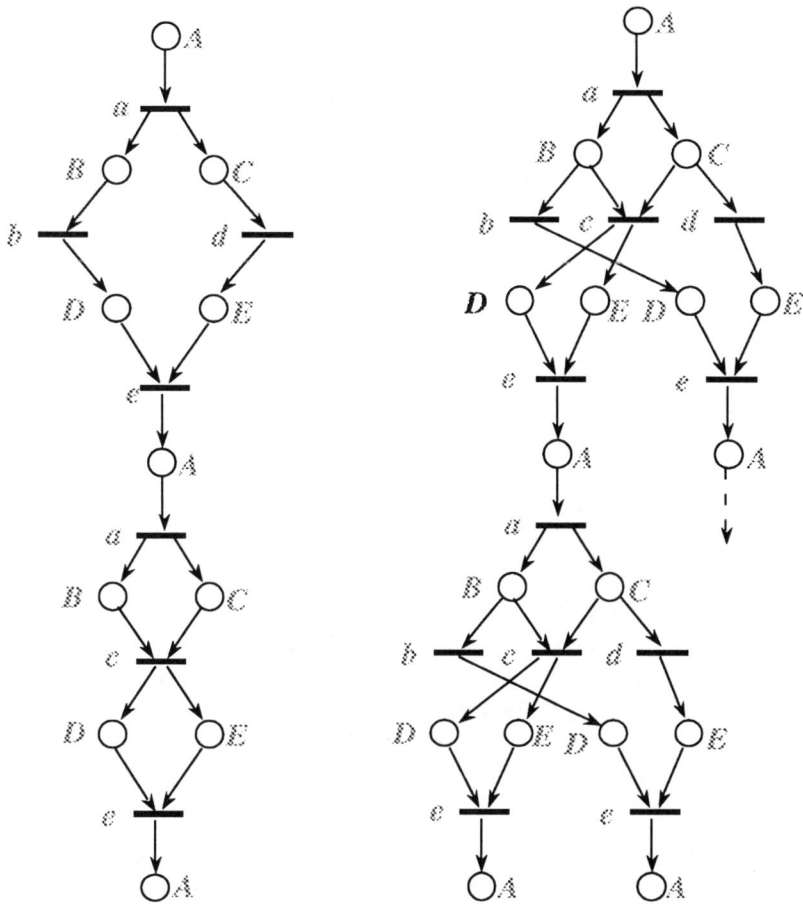

Fig. 3. Deterministic and non-deterministic processes the net of Figure 1

In the right side of Figure 3 we depicted a *non-deterministic process* for the same
net, which is almost like a deterministic processes, but where forward conflicts
are allowed (i.e., there can be many arcs outgoing from the same place). It
is not difficult to construct a non-deterministic process of a given (safe) net
by "unfolding" it, and by duplicating places when needed to avoid backward
conflicts. Also a non-deterministic process enjoys the property that every firing
sequence in it corresponds to a firing sequence of the original net; moreover, the

infinite process obtained by completing in the expected way the one shown in the right side of Figure 3 can be uniquely characterized by the fact that every firing sequence of net N is the image of exactly one firing sequence of the process.

Finally, in Figure 4 the prime event structure corresponding to the non-deterministic process of Figure 3 is shown. We recall here the formal definition of prime event structures: a comprehensive treatment of can be found in [Win87].

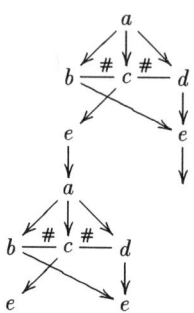

Fig. 4. The prime event structure of the net of Figure 1

Definition 3 (prime event structures). A **prime event structure** $\mathcal{E}$ is a triple $\mathcal{E} = \langle E, \leq, \# \rangle$ where:

- E is a set of events.
- $\leq \,\subseteq E \times E$ is a partial order relation which satisfies the **axiom of finite causes**, i.e., $\{e' \mid e' \leq e\}$ is finite for all $e \in E$. Relation $\leq$ is called the **causal dependency relation**.
- $\# \subseteq E \times E$ is a binary, symmetric, irreflexive relation which is **hereditary**, i.e., if $e \# e_1$ and $e_1 \leq e_2$, then $e \# e_2$. Relation $\#$ is called the **conflict relation**. □

In the following by "event structure" we will always mean a "prime" one. In the event structure of Figure 4 the causal dependency relation is represented by its Hasse diagram with directed arcs, and the conflict relation is drawn as undirected arcs labelled by a "$\#$": only conflicts which cannot be inherited are shown. The event structure is easily obtained from the non-deterministic process by deleting all places, letting $t \leq t'$ if a postcondition of t is a precondition of t', letting $t \# t'$ if t and t' have a common precondition (i.e., if they are in conflict), and closing relation $\leq$ under transitivity and symmetry and $\#$ under inheritance.

2.2 Semantics of nets, categorically

A fundamental result of [Win87] is that the truly-concurrent semantics of safe nets can be given via adjunctions between suitable categories. More precisely, let

Safe be the category of safe nets (we will not report the definition of the arrows of the involved categories: see [Win87] for that), let **Occ** be the category of (non-deterministic) occurrence nets, and let **ES** be the category of event structures. Then **Occ** clearly has an inclusion functor I to **Safe** (every occurrence net is safe); on the other hand, the unfolding procedure that associates an occurrence net with a safe net defines a functor $U : \textbf{Safe} \to \textbf{Occ}$, which is the right adjoint to I: thus **Occ** is a *reflexive subcategory*[3] of **Safe**. Also the construction of the event structure from the occurrence net sketched above can be turned into a functor $ES : \textbf{Occ} \to \textbf{ES}$; moreover there is a functor $O : \textbf{ES} \to \textbf{Occ}$ which generates an occurrence net from an event structure by regarding the events as transitions and by adding places in a suitable way. It can be shown that ES is the right adjoint to O.

$$\textbf{Safe} \xleftarrow[U]{I} \textbf{Occ} \xleftarrow[ES]{O} \textbf{ES} \xleftarrow[L]{Pr} \textbf{Dom}$$

Fig. 5. Chain of adjunctions from safe nets to prime algebraic domains [Win87]

The two adjunctions introduced so far can be concatenated yielding an adjunction between **Safe** and **ES**, as shown in Figure 5. The last adjunction shown in the figure is between event structures and the category **Dom** of *prime algebraic domains*; actually, it is an equivalence of categories, obtained as follows. A *configuration* of a given event structure is a subset of its events which is left-closed (with respect to the causal dependency relation) and conflict-free; the set of all configurations of an event structure $\mathcal{E}$ can be shown to form a prime algebraic domain $L(\mathcal{E})$ (i.e., a Scott domain satisfying some additional properties), when ordered by inclusion. Conversely, the set of "prime" elements of a prime algebraic domain (see [Win87] for definitions) form a prime event structure, and these two mappings can be extended to functors forming an adjunction.

Remark. What's the point in using category theory to relate systems (e.g., nets) with their semantics (e.g., event structures or domains)? Is it not sufficient to give the explicit construction of the semantics for a given system?

There are (at least) three good reasons for using categories:

1. When defining a *category* of systems one is forced to provide a notion of morphism, checking that the axioms of categories are satisfied. Often this procedure gives important insights about the structure of systems. For example, the notion of *isomorphism* is derived by that of morphism, and relates system which are "conceptually" the same: all the categorical constructions will handle isomorphic systems in a uniform way.

---

[3] A subcategory is reflexive if the inclusion functor has a right adjoint; it is coreflexive if the inclusion has a left adjoint.

Moreover, one can check for the existence of some categorical constructions (like products and coproducts, or limits and colimits in general) which should correspond to suitable operations on systems. Performed in a category, such operations are in general not deterministic (limits and colimits are unique only up to isomorphisms), but this apparent drawback turns out in many situations to be a real simplification. Indeed, thinking for example to operations which glue together (parts of) systems, all the (syntactical) problems related with naming (like α-conversion in logical systems, the "renaming apart" of variables in logic programming, the choice of new names when building the disjoint union of systems) simply disappear in the categorical framework: such operations often correspond to colimit constructions (see also [Gog91]).

2. Once a category of systems and one of "denotations" (semantics) are defined, there are in general many ways (if any) to map the first ones to the others and viceversa. The categorical framework forces you to define these mappings in a consistent way on morphisms as well (because they have to be functors).

3. There are in general many pairs of functors relating two categories. Nevertheless, often (but by no mean always) given two "related" categories (of systems, denotations or whatever) there happen to be an "obvious" functor in one direction. See for example, for the categories above, the inclusion of **Occ** into **Safe**, or the functor $ES : \textbf{Occ} \to \textbf{ES}$. Keeping such functors fixed, one can look for functors in the opposite direction forming an adjunction: if such a functor exists it is unique (up to a natural transformation, by general categorical results). The fact that two functors form an adjunction is often regarded as a good argument in favour of the "correctness" and "naturalness" of the relationship established between two categories.

The chain of adjunctions shown in Figure 5 is just a prototypical example of a general technique in the categorical semantics of concurrency. Other adjuctions relating categories of systems can be found in [Bed88, MMS92, SNW93].

3 From Petri nets to graph grammars

Although the state of a Place/Transition Petri net has an intrinsically distributed nature, its structure is undoubtedly poor. Formally, it is just a multiset of places, or equivalently an isomorphism class of sets (of tokens) labelled by places. Thus a P/T net is just a (multi)set rewrite system.

Even if this is sufficient for many purposes, it is easy to believe that in more complex situations the state of a distributed system cannot be described faithfully as a set of unstructured components. One possible generalization of such nets is obtained by adding structure to the tokens: this idea is at the basis of the definition of *High-Level Petri nets*, including for example Colored and Predicate/Transition nets (see [JR92], and also the contribution by Kurt Jensen in this volume). Alternatively, one might be interested to represent explicitly also the *relationships* among different components of a distributed state: thus graphs may turn out to be much more expressive than multisets for representing such states. For example, there are obvious natural graphical representations of systems of processes interacting through channels, of predicates acting on shared variables (e.g., in logic programming), and of the various kind of relationships among the files of a file system, or among the classes of records of a database system, or among the components of a complex software system, and so on.

A direct consequence of these considerations is that formalisms based on graph rewriting must be more expressive than P/T Petri nets for specifying the evolution of systems whose states have a distributed and interconnected structure. After introducing the basics of graph grammars, we will substantiate this claim showing in which sense they properly extend P/T nets.

Clearly, the question arises whether graph grammars as well can be equipped with computation-based semantics, and possibly through clean categorical techniques (as shown for nets in Sections 2.1 and 2.2). We will address this point in Section 4 summarizing recent results appeared in the literature and pointing to some ongoing research activities.

The *theory of graph grammars* (or of *graph rewriting*) originated in the late 60s, and basically studies a variety of formalisms which extend the theory of formal languages in order to deal with structures more general than strings, like graphs and maps. A graph grammar allows one to describe finitely a (possibly infinite) collection of graphs, i.e., those graphs which can be obtained from a start graph through repeated applications of graph productions. Each production can be applied to a graph by replacing an occurrence of its left-hand side with its right-hand side. The form of graph productions and the mechanisms stating how a production can be applied to a graph and what the resulting graph is, depend on the specific formalism.

Graph grammars have been applied in different areas of computer science: among them we recall data bases, software specification, incremental compilers and pattern recognition (see [CER79, ENR83, ENRR87, EKR91], the proceedings of the first four international workshops on graph grammars). They have also been used in various ways for the specification of concurrent and distributed systems, like actor systems (e.g., in [JR89]), Petri nets (e.g., in [GJRT83]), and systems of distributed processes [DM87]). However, only recently [CELMR94b] they have been recognized as a very natural generalization of Petri nets, in the way they are presented in this paper.

Among the various formulations of graph rewriting, the so called "algebraic approach" [Ehr87] is one of the most successful, mainly because of its flexibility. In fact, since the basic notions of production and direct derivation are defined in terms of diagrams and constructions in a category, they can be applied in a uniform way to a wide range of structures, simply by changing the underlying category. This generality of the approach has been exploited for example in the definition of *attributed graph grammars* [LKW93] and of the so-called *High-Level Replacement Systems* [EHKP91]. It is not difficult to imagine that the strong relationship between P/T nets and the rewriting of directed, labeled graphs explored below in Section 3.2 could be extended by considering high-level nets on one side, and suitable kinds of attributed grammars or of high-level replacement systems on the other side.

3.1 Basic definitions

There exist many different notions of graphs: one main advantage of the algebraic approach is that most definitions can be given once and for all, then they can be applied to different categories of graphs (or of other structures). Thus, although we fix a specific kind of graphs for relating graph grammars to Petri nets, everything can be rephrased for other kinds of graphs.

Definition 4 (colored graphs [Ehr87]). Given two fixed alphabets Ω_E and Ω_V for edge and vertex colors, respectively, a **(colored) graph** (over (Ω_E, Ω_V)) is a tuple $G = \langle E, V, s, t, l_E, l_V \rangle$, where E is a finite set of **edges**, V is a finite set of **vertices**, $s, t : E \to V$ are the **source** and **target** functions, and $l_E : E \to \Omega_E$ and $l_V : V \to \Omega_V$ are the **edge** and the **vertex coloring** functions, respectively. A graph is **discrete** if it has no edges. A **graph morphism** $f : G \to G'$ is a pair of functions $f = (f_E : E \to E', f_V : V \to V')$ which preserve sources, targets, and colors, i.e., such that $f_V \circ t = t' \circ f_E$, $f_V \circ s = s' \circ f_E$, $l'_V \circ f_V = l_V$, and $l'_E \circ f_E = l_E$. A graph morphism f is an **isomorphism** if both f_E and f_V are bijections. If there exists an isomorphism from graph G to graph H, then we write $G \equiv H$; moreover, $[G]$ denotes the isomorphism class of G, i.e., $[G] = \{H \mid H \equiv G\}$. An **automorphism** h on a graph G is an isomorphism $h : G \to G$; an automorphism different from the identity is called **non-trivial**.

The category having colored graphs over (Ω_E, Ω_V) as objects and graph morphisms as arrow is called (Ω_E, Ω_V)-**Graph**, or simply **Graph** if the alphabets of colors are understood. □

The categorical construction most extensively used in the algebraic theory of graph grammars is that of *pushout*. Intuitively, regarding graphs as distributed systems, a pushout specifies how to merge together two systems having a common subsystem [Gog91]. In a set-theoretical framework one should handle explicitly the possibility of clashing of names used in the two systems (in the non-shared parts): this is given for free in the categorical construction, at the price of determining the result only up to isomorphism.

Definition 5 (pushout [ML71] and pushout complement [Ehr87]). Given a category **C** and two arrows $b : A \to B$, $c : A \to C$ of **C**, a triple $\langle D, g : B \to D, f : C \to D \rangle$ as in Figure 6 (a) is called a **pushout** of $\langle b, c \rangle$ if *[Commutativity]* $g \circ b = f \circ c$, and *[Universal Property]* for all objects D' and arrows $g' : B \to D'$ and $f' : C \to D'$, with $g' \circ b = f' \circ c$, there exists a unique arrow $h : D \to D'$ such that $h \circ g = g'$ and $h \circ f = f'$.

In this situation, D is called a **pushout object** of $\langle b, c \rangle$. Moreover, given arrows $b : A \to B$ and $g : B \to D$, a **pushout complement** of $\langle b, g \rangle$ is a triple $\langle C, c : A \to C, f : C \to D \rangle$ such that $\langle D, g, f \rangle$ is a pushout of b and c. In this case C is called a **pushout complement object** of $\langle b, g \rangle$. □

Definition 6 (graph grammars, direct derivations [Ehr87]). A (graph) **production** $p = (L \xleftarrow{l} K \xrightarrow{r} R)$ is a pair of injective graph morphisms $l : K \to L$

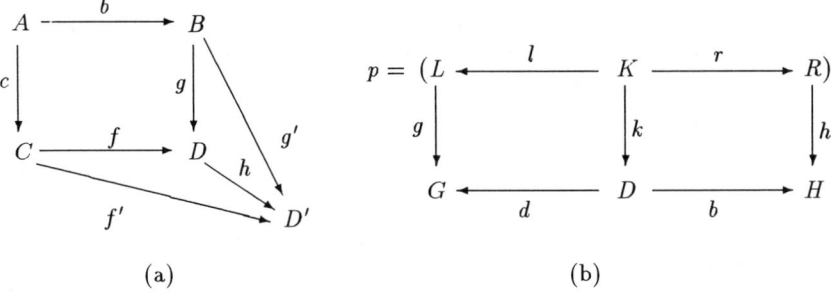

Fig. 6. (a) Pushout diagram. (b) Direct derivation as double-pushout construction.

and $r : K \to R$. The graphs L, K, and R are called the **left-hand side**, the **interface**, and the **right-hand side** of p, respectively. Production p is called **consumptive** if morphism $l : K \to L$ is not surjective. A **graph grammar** $\mathcal{G} = (\{p_i\}_{i \in I}, \underline{G})$ is a set of graph productions together with a **start graph** $\underline{G}$.

Given a graph G, a graph production $p = (L \xleftarrow{l} K \xrightarrow{r} R)$, and an **occurrence** (i.e., a graph morphism) $g : L \to G$, a **direct derivation** α **from** G **to** H **using** p **(based on** g**)** exists if and only if the diagram in Figure 6 (b) can be constructed, where both squares are required to be pushouts in **Graph**. In this case, D is called the **context** graph, and we write $\alpha : G \Rightarrow_{p,g} H$, or simply $\alpha : G \Rightarrow_p H$. □

In a graph-theoretical setting, the pushout object D of Figure 6 (a) can be understood as the gluing of graphs B and C, obtained by identifying the images of A along b and c. Therefore the double-pushout construction can be interpreted as follows. In order to apply the production p to G, we first need to find an occurrence of its left-hand side L in G, i.e., a graph morphism $g : L \to G$. Next, to model the deletion of that occurrence from G, we have to find a graph D and morphisms k and d such that the resulting square is a pushout: The context graph D is therefore required to be a pushout complement object of $\langle l, g \rangle$. Finally, we have to embed the right-hand side R into D: This embedding is expressed by the right pushout.

It is worth stressing that category **Graph** has all pushouts, but that the pushout complement of two arrows does not always exist, and if it exists it is not necessarily unique: Conditions for the existence (called *gluing conditions*) are presented in [Ehr87], and in our specific case the assumption that morphism l is injective guarantees that if the pushout complement exists then it is unique (up to isomorphisms).

3.2 Petri nets as graph grammars

Since Petri nets are "(multi)set rewriting systems" and sets can be regarded as discrete graphs, it is easy to believe that nets can be regarded as graph grammars satisfying some additional constraints. This relationship is formalized here, by showing how a P/T net can be transformed into a graph grammar. The idea is to regard markings as discrete graphs colored with places, and transitions as graph productions. Therefore in the rest of this section we assume to work in category (Ω_E, S)-**Graph**, including all colored graphs over (Ω_E, S); in other words, the colors for vertices are exactly the places of **N**.

Definition 7 (the graph grammar associated with a P/T net). Let $\mathbf{N} = \langle S, T, F, W, \underline{M} \rangle$ be a P/T net. If $M : S \to \mathbb{N}$ is a marking for **N**, then the **graph associated with** M is the discrete graph $\partial(M) = \langle \emptyset, V_{\partial(M)}, \emptyset, \emptyset, \emptyset, l_{V_{\partial(M)}} \rangle$ where $V_{\partial(M)} = \{\langle s, i \rangle \mid s \in S \wedge 1 \leq i \leq M(s)\}$, and $l_{V_{\partial(M)}}(\langle s, i \rangle) = s$ for all $\langle s, i \rangle \in V_{\partial(M)}$. Conversely, the **marking induced by a discrete graph** $G = \langle \emptyset, V_G, \emptyset, \emptyset, \emptyset, l_{V_G} \rangle$ is defined as $\mu(G)(s) = \#\{v \in V_G \mid l_{V_G}(v) = s\}$ for all $s \in S$. Notice that $\mu(\partial(M)) = M$, and that $G \equiv H$ implies $\mu(G) = \mu(H)$, because graph morphisms must preserve colors.

The **graph grammar associated with N** is the graph grammar $\mathbf{GG}(\mathbf{N}) = (\{p_t\}_{t \in T}, \underline{G})$, where the start graph is $\underline{G} = \partial(\underline{M})$, and for each $t \in T$ production $p_t = \left(L \xleftarrow{l} K \xrightarrow{r} R \right)$ is defined in terms of transition t as follows:

- $L = \partial(pre(t))$.
- K is the empty graph and l, r are the empty graph morphisms.
- $R = \partial(post(t))$. □

The last definition formalizes the fact that P/T nets are a proper subclass of graph grammars. In fact, they correspond to grammars acting on discrete graphs only, such that the interface graph is empty in all productions. The next result shows that transforming a net into a graph grammars preserves its dynamic behaviour.

Proposition 8 (preservation of behaviour). *Let M be a marking for a P/T net **N** and suppose that $M[t\rangle M'$ (i.e., t is enabled at M and the result of its firing is M'). Then there is a direct derivation step $\partial(M) \Rightarrow_{p_t, g} H$ using the production associated with t. Moreover, we have that $\mu(H) = M'$. Conversely, if G is discrete and $G \Rightarrow_{p_t, g} H$, then t is enabled at $\mu(G)$ and $\mu(G)[t\rangle\mu(H)$.* □

The correspondence between nets and grammars can be specialized to the subclasses of safe systems as follows. Safe nets have been informally introduced just before Section 2.1.

Definition 9 (safe graph grammars). A graph grammar $\mathcal{G} = (\{p_i\}_{i \in I}, \underline{G})$ is **safe** if (1) every production in $\mathcal{G}$ is consumptive, and (2) for every concrete derivation $\rho : \underline{G} \Rightarrow^* H$ starting from the start graph $\underline{G}$, graph H has no non-trivial automorphisms. □

Proposition 10 (safe nets and safe grammars). *Let **N** be a P/T net with no isolated transitions (t is isolated if $pre(t) = post(t) = 0$), and let $\mathbf{GG}(\mathbf{N})$ be its associated graph grammar. Then **N** is safe iff $\mathbf{GG}(\mathbf{N})$ is safe.* □

4 Towards a concurrent semantics for graph grammars

The algebraic theory of graph grammars comprises many results concerning parallelism and concurrency (see [Kre77, Kre87, Ehr87]). Most of those results recast, in the more general framework of graph rewriting, related notions and results of (term) rewriting systems, exploring properties of confluence, Church-Rosser, orthogonality of redexes, parallel moves, and so on. Thus they are mainly concerned with the study of properties of derivations and of their syntactic manipulations. Such aspects of a graph grammar are surely of great interest. However, in the perspective of presenting graph grammars as a more adequate formalism than nets for the specification of concurrent/distributed systems, also (possibly categorical) truly-concurrent semantics like those sketched in Section 2 for Petri nets would be interesting.

The development of such semantics if the main topic of a joint research activity of the authors with Hartmut Ehrig and Michael Löwe from Berlin and with Francesca Rossi from Pisa. As a matter of fact, the results about nets cannot be generalized straightforwardly for graph grammars, because they extend Petri nets with some non-trivial features. In fact, in Section 3.2 we showed that graph grammars extend Petri nets not only because they act on graphs instead of on (multi)sets, but also because graph productions in general have a non-empty interface, unlike nets. This makes possibile to specify a sort of context-dependent rewriting, i.e., the fact that the application of a production does not have to consume all its preconditions, but just part of them. For example, the rule "a rewrites to b if c is present" can be represented by a production of the form $R_1 : \{a, c\} \leftarrow \{c\} \rightarrow \{b, c\}$, where the non-empty interface $\{c\}$ specifies that c is preserved. On the other hand, the rule "a and c rewrite to b and c" is represented by the production $R_2 : \{a, c\} \leftarrow \emptyset \rightarrow \{b, c\}$, where $\emptyset$ is the empty graph. This feature can be used to specify faithfully any kind of "read" operation, and it is not present in nets. In fact in a Petri net only production R_2 has a natural representation as a transition which consumes two tokens (from places a and c), and produces two tokens (in places b and c), while the effect of production R_1 must be simulated by the same transition. It turns out that this context-dependent aspect of graph rewriting is even more difficult to be treated formally than the generalization from sets to graphs. Recently an extension of the Petri net model to include a notion of context-dependent rewriting has been proposed in [MR93].

To give an idea about the results obtained so far and those we are looking for, we summarize in Figure 7 the relationships among various papers.

In [CELMR94a] we addressed the problem of giving a correct notion of "abstract derivations" for a graph grammar, able to abstract out from unnecessary details, but still allowing for sequential composition. The main point was that a naive definition based on isomorphism classes of derivations did not work, but a more sophisticated solution was needed. An interesting result was the definition of a *category of abstract derivations* of a grammar, i.e., a category having abstract graphs as objects and graph derivations as arrows, quotiented with respect to an equivalence relation which relates derivations differing only for representation

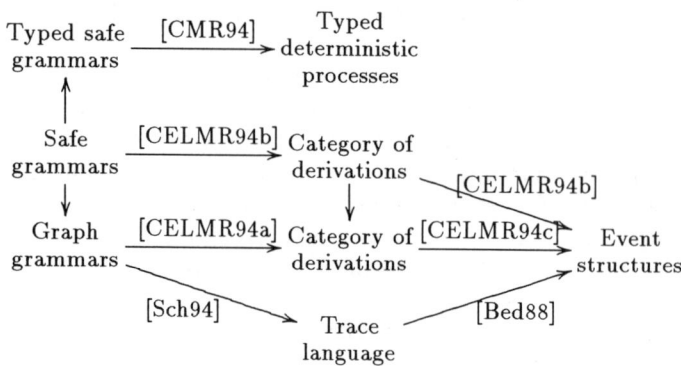

Fig. 7. Recent and ongoing works on truly-concurrent semantics for graph grammars

details or for the order in which "independent" direct derivations are performed. This category provides, in our view, a very adequate *model of computation* for a graph grammar. For the definition of such a model we were inspired by the definition of the category $P[N]$ of a Petri net N, as proposed in [DMM89]; such a category has markings as objects, and as arrows deterministic "concatenable" processes (a slight variation of the processes introduced in Section 2.1). Indeed, if the construction of the category of derivation proposed in [CELMR94a] is applied to the graph grammar associated with a net N, the resulting category is closely related to $P[N]$. The paper also made evident that restricting to the class of safe grammars the construction of such a category is much easier.

Our most relevant result, for the time being, is the definition of an event structure semantics for the subclass of safe grammars [CELMR94b]. An important point is that the construction of the event structure of a grammar we proposed is quite simple, as it does not go through the definition and construction of a non-deterministic process, as shown for nets in Section 2.1. Instead, we take the comma category of the objects under the initial graph in the category of abstract derivations of the grammar, and this comma category is proved to be a prime algebraic domain; thus a prime event structure can be extracted from it, thanks to the last adjunction of Figure 5. It is worth stressing that if the proposed construction is applied to the safe graph grammar representing a safe Petri net (see Definition 7 and Proposition 10), then the resulting event structure is equal to that obtained by the construction proposed in [NPW81, Win87]. In the work in progress [CELMR94c] we are extending the construction of the event structure to arbitrary consumptive grammars.

We are also interested in understanding how the notion of nonsequential process can be generalized from nets to grammars. A first contribution in this line is [CMR94] where we considered just deterministic processes and safe grammars. Actually, in this paper we introduced a slight generalization of grammars, called *typed graph grammars*: These are standard grammars where the rewritten graphs have a morphism to a fixed *type graph*, and turned out to be useful also for the definition of a category of graph grammars, as proposed in [CELMP94a].

A paper closely related to the research activity described so far is [Sch94], where Georg Schied shows how to construct a prime event structure from an arbitrary graph grammar. His approach substantially differs from ours for two main reasons, although the results are similar. First, he uses a different technique to get to the event structure, constructing as an intermediate step a *trace language* and then applying general results from [Bed88]. Second, he uses a more set-theoretical definition of graph rewriting, where the result of a direct derivation is uniquely determined (not up to isomorphisms as in our case).

In conclusion, we explained why graph grammars are in our view a very interesting generalization of P/T Petri nets, presenting the formal relationship between the two formalisms. We also stressed that still much work has to be done to provide graph grammars with a truly concurrent semantics comparable to that of Petri nets, and that this area of research is quickly evolving.

References

[Bed88] M.A. Bednarczyk, *Categories of asynchronous systems*, Ph.D. Thesis, University of Sussex, Report no. 1/88, 1988.

[CER79] V. Claus, H. Ehrig and G. Rozenberg (Eds.), *Proceedings of the 1st International Workshop on Graph-Grammars and Their Application to Computer Science and Biology*, LNCS 73, Springer-Verlag, 1979.

[CELMR94a] A. Corradini, H. Ehrig, M. Löwe, U. Montanari and F. Rossi, *Abstract Graph Derivations in the Double-Pushout Approach*, in [SE94], 1994.

[CELMR94b] A. Corradini, H. Ehrig, M. Löwe, U. Montanari and F. Rossi, *An event structure semantics for safe graph grammars*, to appear in *Proceedings of the IFIP Working Conference PROCOMET '94*.

[CELMR94c] A. Corradini, H. Ehrig, M. Löwe, U. Montanari and F. Rossi, *An Event Structure Semantics for Consumptive Graph Grammars with Parallel Productions*, extended abstract, 1994.

[CELMP94a] A. Corradini, H. Ehrig, M. Löwe, U. Montanari and J. Padberg, *Typed Graph Grammars and their Adjunction with Categories of Derivations*, extended abstract, 1994.

[CELMP94b] A. Corradini, H. Ehrig, M. Löwe, U. Montanari and J. Padberg, *Functorial Semantics for Safe Graph Grammars Using Prime Algebraic Domains*, extended abstract, 1994.

[CMR94] A. Corradini, U. Montanari and F. Rossi, *Graph Processes*, submitted for publication.

[DMM89] P. Degano, J. Meseguer, U. Montanari, *Axiomatizing Net Computations and Processes*, in Proc. 4th Annual Symposium on Logic in Computer Science, Asilomar, CA, USA, 1989, 175–185.

[DM87] P. Degano and U. Montanari, *A model of distributed systems based on graph rewriting*, in *Journal of the ACM* **34** (1987), 411–449.

[Ehr87] H. Ehrig, *Tutorial introduction to the algebraic approach of graph-grammars*, in [ENRR87], 3–14.

[EHKP91] H. Ehrig, A. Habel, H.-J. Kreowski and F. Parisi-Presicce, *Parallelism and Concurrency in High-Level Replacement Systems*, in Mathematical Structures in Computer Science **1** (1991), 361–404.

[EKR91] H. Ehrig, H.-J. Kreowski and G. Rozenberg (Eds.), *Proceedings of the 4th International Workshop on Graph-Grammars and Their Application to Computer Science*, LNCS 532, Springer-Verlag, 1991.

[ENR83] H. Ehrig, M. Nagl and G. Rozenberg (Eds.), *Proceedings of the 2nd International Workshop on Graph-Grammars and Their Application to Computer Science*, LNCS 153, Springer-Verlag, 1983.

[ENRR87] H. Ehrig, M. Nagl, G. Rozenberg and A. Rosenfeld (Eds.), *Proceedings of the 3rd International Workshop on Graph-Grammars and Their Application to Computer Science*, LNCS 291, Springer-Verlag, 1987.

[Gog91] J.A. Goguen, *A categorical manifesto*, Math. Struc. Comput. Sci. 1(1991).

[GJRT83] H.J. Genrich, D. Janssens, G. Rozenberg and P.S. Thiagarajan, *Petri nets and their relation to graph grammars*, in [ENR83], 115–129.

[JR89] D. Janssens and G. Rozenberg, *Actor Systems*, in Math. Systems Theory **22** (1989), 75–107.

[JR92] K. Jensen and G. Rozenberg (Eds.), *High-level Petri nets: theory and application*, Springer-Verlag, Berlin, 1992.

[Kre77] H.-J. Kreowski, *Manipulation von Graph Transformationen*, Ph.D. Thesis, Technische Universität Berlin, 1977.

[Kre87] H.-J. Kreowski, *Is parallelism already concurrency? Part 1: Derivations in graph grammars*, in [ENRR87], 343–360.

[LKW93] M. Löwe, M. Korff, and A. Wagner, *An Algebraic Framework for the Transformation of Attributed Graphs*, in M.R. Sleep, M.J. Plasmeijer, and M.C. van Eekelen (Eds.), *Term Graph Rewriting: Theory and Practice*, Wiley, London, 1993, 185–199.

[ML71] S. Mac Lane, *Categories for the working mathematician*, Springer, 1971.

[MMS92] J. Meseguer, U. Montanari, and V. Sassone, *On the semantics of Petri Nets*, in Proceedings CONCUR '92, Springer-Verlag, LNCS 630, 286–301.

[MR93] U. Montanari and F. Rossi, *Contextual nets*, TR 4-93, Dipartimento di Informatica, University of Pisa, Italy, February 1993.

[NPW81] M. Nielsen, G. Plotkin and G. Winskel, *Petri Nets, Event Structures and Domains, Part 1*, in Theoret. Comput. Sci. **13** (1981), 85–108.

[Pet62] C.A. Petri, *Kommunikation mit Automaten*, Schriften des Institutes für Instrumentelle Matematik, Bonn, 1962.

[Rei85] W. Reisig, *Petri Nets: An Introduction*, EACTS Monographs on Theoretical Computer Science, Springer-Verlag, 1985.

[Sch94] G. Schied, *On relating Rewriting Systems and Graph Grammars to Event Structures*, in [SE94], 326–340.

[SE94] H.J. Schneider and H. Ehrig (Eds.), *Proceedings of the Dagstuhl Seminar 9301 on Graph Transformations in Computer Science*, LNCS 776, Springer-Verlag, 1994.

[SNW93] V. Sassone, M. Nielsen and G. Winskel, *Relationships between models of concurrency*, in Proceedings REX '93, 1993.

[Win87] G. Winskel, *Event Structures*, in Petri Nets: Applications and Relationships to Other Models of Concurrency, LNCS 255, Springer-Verlag, 1987, 325–392.

Towards a Theory of Strong Bisimulation for the Service Rendezvous

Michael Baldamus

Technische Universität Berlin, Institut für Software und Theoretische
Informatik, Sekretariat FR 6–10, Franklinstraße 28/29, 10587 Berlin, Germany
e–mail: baldamus@cs.tu-berlin.de

Abstract. In the present paper we start from an operational framework of concurrent processes in which the discipline of process interaction is the service rendezvous. Our objective consists in examining two notions of rendezvous based strong bisimulation equivalence. The first one is quite immediate and we can prove that it is a congruence. The second one is slightly less immediate but for it we can give a neat alternative characterization in terms of a minor variant of Aczel's ε–bisimulation.

Our motivation for using ε–bisimulation stems from the presence of conceptual non–well–foundedness in our framework. What is crucial, we bring this form of bisimulation to bear not in a set theoretical but in a logical context. Despite conceptual non–well–foundedness we, thus, do not have to leave standard set theory.

The conceptual non–well–foundedness is induced by a (very limited) higher order capability. To overcome certain technical difficulties arising in connection with it we transfer methods developed by Thomsen in the context of his Plain CHOCS higher order process calculus. In neither direction, however, a meaningful embedding seems to be possible.

1 Introduction

In process algebra a whole sphere of different notions of bisimulation has emerged. To structure this sphere the following distinctions can be made:

(1) as to weak and strong forms of bisimulation.
(2) as to the underlying process interaction discipline. Examples are notions of bisimulation based on pure synchronization (see, e.g., [Mil89]), value passing (see, e.g., [HL93]), label passing (see, e.g., [MP89]), and higher order processes (see, e.g., [Tho93]).

Here we want to introduce two notions of strong bisimulation that are based on the service rendezvous (rendezvous) between client and server processes (clients

---

The research for this work has been supported by the Deutsche Forschungsgemeinschaft (German Research Society) under grant Ho 1257/2.

and servers). We proceed as follows:

- In Section 2 we lay the groundwork by defining a language and a Structural Operational Semantics (SOS) of that language. Both together do the modeling task for us. In particular, the rendezvous appears as in many programming languages where it is built–in, namely as a primitive. Apart from the examples Section 2 actually represents a rerun of the corresponding parts of [Bal94]. Hence it is rather succinct and does not go too far into the rationale of our framework.

- In Section 3 we introduce the two notions of strong bisimulation mentioned above. The first one is derived in an immediate way from the language and its SOS. The second one is less immediate. What is interesting, it is possible to give an alternative characterization of the equivalence emerging from it that one expects but has not been able to prove or disprove in the case of the first notion.
The main result in Section 3 says that the equivalence emerging from our first notion is a congruence.

- In Section 4 we give the alternative characterization just mentioned. It is stated in terms of so called i–ε–bisimulation equivalence over the set of objects of a so called i–ε–structure. These two are minor variants of the concepts of ε–bisimulation equivalence and ε–structure well known from the theory of non–well–founded sets (cf. [Acz88]) or set theory in general, respectively. Following an idea by Mahr et al. (cf. [MSU90]) we employ them to simulate a semantic domain for our language. Otherwise we would have to retort to non–well–founded sets, as in our model of the rendezvous we make essential use of functions which are indirectly self–applicable.

- In Section 5 we give some concluding remarks.

- In an appendix we give an outline of the proof of the above mentioned congruence property.

Related Work

There seems to be no published work in which notions of rendezvous based bisimulation appear (cf. [BZ83], [Bak89], [BV91a], [BV91b], [Egg91], [Fra94a], and [Fra94b]). In the case of [Fra94a], though, one might speak of hidden ones. The reason is that in this work Frauenstein describes a π–calculus semantics of a language in whose design the rendezvous plays a central role in bringing about a combination of concepts from applicative and process oriented programming (an idea originally developed by Egger in [Egg91]); and the preimage of any bisimulation over the π–calculus might reasonably be defined to be a bisimulation of the same type over that language. Such, however, has not been made explicit or analyzed.

In fact, as far as the rendezvous is concerned most of the works cited above more or less do without any intermediate language; [BZ83] and [Fra94a] are the only exceptions.

Acknowledgement. For discussions during the preparation of the paper the author is grateful to Rocco De Nicola, Bernd Mahr, Thomas Frauenstein, Rainer Glas, and Philip Zeitz. Thanks also to Hartmut Ehrig for suggesting some final improvements.

2 Preliminaries

In programming languages such as Ada ([DoD83]) or Egger's AP ([Egg91]) every particular rendezvous comprises an explicit synchronization at its beginning and an implicit one at its end. By the first one the client enters a wait–state and by the second one it leaves that state, the service being everything done by the server in between these two events. In [Bal94] this discipline is traced back to a kind of continuation passing mechanism: First, requests and offers for services are viewed as uninterpreted actions. Second, if a synchronization between an offer and a request takes place then the residual of the client is passed to the residual of the server, i.e. the latter is a function of type process $\longrightarrow$ process. The result is a process which is about to execute the service and which starts the residual of the client in parallel with the post–service activity of the server thereafter. Technically this is achieved by letting the client residual become unguarded. As a consequence the second synchronization between the client and the server is indeed implicit. In this sense we can speak of a rendezvous primitive in our framework.

Here the following syntax is our starting point for implementing this mechanism:

$$T ::= \text{nil} \mid a \to T \mid \bar{a} \to \lambda X.T \mid \tau \to T \mid T + T \mid T|T \mid T \backslash a \mid X.$$

It is based on a set *Name* of *action names*, ranged over by $a, b, \ldots$, and a set of *variables*, ranged over by $X, Y, \ldots$, which are assumed to be given and must be at least countably infinite. Besides that the set $\overline{Name}$ of *complementary action names*, ranged over by $\bar{a}, \ldots$, is defined by $\overline{Name} = \{\bar{a} : a \in Name\}$. Action prefixes of the form $a \to T$ or $\bar{a} \to \lambda X.T$ specify a request or an offer, respectively, for a service. (The reader may note that in accordance to what is explained above an offer takes a function as its second argument.) Action prefixes of the form $\tau \to T$, the other operators, and the constant nil have their usual informal meaning. The set of all expressions is ranged over by T, U and the set of all expressions that are closed wrt. λ–abstraction, called the set of *processes*, is denoted by Pr and ranged over by $P, Q, \ldots$. For every X the set $\Pr(X)$ is defined by $\Pr(X) = \{T : \text{fv}(T) \subseteq \{X\}\}$ where, for every T, $\text{fv}(T)$ denotes the set of all variables that occur unbound in T.

In comparison to [Bal94] the main difference lies in the modeling of internal nondeterminism: While in [Bal94] two choice operators are employed (internal and external choice) we have a CCS–like choice operator and a CCS–like τ–action here (cf. [Mil89]). Besides that the language contains no operator for recursion. (See, however, the example below.)

To the language Pr we assign an operational semantics. It consists of the following three transitional relations:

$\longrightarrow\ \subseteq\ \text{Pr} \times \textit{Name} \times (\text{Pr} \times \text{Pr} \times \mathscr{P}_{\text{fin}}(\textit{Name}))$,
called the relation of *request transitions* (*r–transitions*),

$\longrightarrow\ \subseteq\ \text{Pr} \times \overline{\textit{Name}} \times \{\lambda X.T \mid T \in \text{Pr}(X)\}$,
called the relation of *offer transitions* (*o–transitions*), and

$\overset{\tau}{\longrightarrow}\ \subseteq\ \text{Pr} \times \text{Pr}$,
called the relation of *invisible transitions* (τ–*transitions*).

They are defined to be the least ones that obey the structural axioms and rules listed in Table 1. Except for the CCS–like handling of internal nondeterminism everything is quite similar to the corresponding material in [Bal94]. Thus, let us just note that A stands for an element of the set $\mathscr{P}_{\text{fin}}(\textit{Name})$ and that $\backslash A$ is an abbreviation for $\backslash a_1 \ldots \backslash a_k$, given that $A = \{a_1, \ldots, a_k\}$ where $k \geq 0$. Moreover, in every expression of the form $T\backslash a$ any occurrence of a in an action prefix is *bound* by the restriction if it is not already bound by some restriction in T. The set of all action names occurring unbound in some expression T is denoted by $\text{fn}(T)$. Expressions are regarded as equal if they are syntactically equal up to α–conversion of bound variables and action names and, finally, whenever we substitute an expression for a variable then we implicitly use α–conversion so as to prevent any capture of free variables or action names. — To do α–conversion of action names is actually necessary in order to obtain a proper model of the rendezvous. For details see [Bal94].

The following example gives the reason why our language contains no recursion operator.

1 Example. (A Limited Form of Recursion) To describe recursion we can use a context introduced by Thomsen in connection with his Plain CHOCS higher order process calculus. In Plain CHOCS syntax it has the form

$$((a?X.([]\mid a!X.\text{nil}))\mid a!(a?X.([]\mid a!X.\text{nil})).\text{nil})\backslash a$$

where X is the recursion variable (cf. [Tho93]); here it becomes

$$Y_X[\,] = ((\overline{a}\to\lambda X.([]\mid a\to X))\mid a\to\overline{a}\to\lambda X.([]\mid a\to X))\backslash a.$$

The recursion works because, for every $T \in \text{Pr}(X)$,

$$Y_X[T] \overset{\tau}{\longrightarrow} (T[\overline{a}\to\lambda X.(T\mid a\to X)/X]\mid a\to\overline{a}\to\lambda X.(T\mid a\to X))\backslash a.$$

However, it is a limited form of recursion. The reason becomes clear if we consider the process emanating from the τ–transition above. In this process, there is only one possibility for a synchronization between some copy of the term $\overline{a}\to\lambda X.(T\mid a\to X)$ to the left and the term to the right of the outermost parallel operator. I.e. if there are any parallel copies of the term $\overline{a}\to\lambda X.(T\mid a\to X)$ then at most one of them will ever be put to work.

$a \to P \xrightarrow{a} (P, \text{nil}, \emptyset)$

$P \xrightarrow{a} (P', R, A)$ implies $P + Q \xrightarrow{a} (P', R, A)$ and $Q + P \xrightarrow{a} (P', R, A)$

$P \xrightarrow{a} (P', R, A)$ implies $P|Q \xrightarrow{a} (P', R|Q, A)$ and $Q|P \xrightarrow{a} (P', Q|R, A)$
provided that (*) $A \cap \text{fn}(Q) = \emptyset$

$P \xrightarrow{a} (P', R, A)$ implies $P \backslash b \xrightarrow{a} (P', R, A \cup \{b\})$
provided that $a \neq b$ and (*) $b \notin A$

$\bar{a} \to \lambda X.T \xrightarrow{\bar{a}} \lambda X.T$

$P \xrightarrow{\bar{a}} \lambda X.T$ implies $P + Q \xrightarrow{\bar{a}} \lambda X.T$ and $Q + P \xrightarrow{\bar{a}} \lambda X.T$

$P \xrightarrow{\bar{a}} \lambda X.T$ implies $P|Q \xrightarrow{\bar{a}} \lambda X.(T|Q)$ and $Q|P \xrightarrow{\bar{a}} \lambda X.(Q|T)$

$P \xrightarrow{\bar{a}} \lambda X.T$ implies $P \backslash b \xrightarrow{\bar{a}} \lambda X.(T \backslash b)$ provided that $a \neq b$

$\tau \to P \xrightarrow{\tau} P$

$P \xrightarrow{\tau} P'$ implies $P + Q \xrightarrow{\tau} P'$ and $Q + P \xrightarrow{\tau} P'$

$P \xrightarrow{\tau} P'$ implies $P|Q \xrightarrow{\tau} P'|Q$ and $Q|P \xrightarrow{\tau} Q|P'$

$P \xrightarrow{a} (P', R, A)$ and $Q \xrightarrow{\bar{a}} \lambda X.T$ implies
$P|Q \xrightarrow{\tau} (R|T[P'/X]) \backslash A$ and $Q|P \xrightarrow{\tau} (T[P'/X]|R) \backslash A$
provided that (*) $A \cap \text{fn}(T) = \emptyset$

$P \xrightarrow{\tau} P'$ implies $P \backslash a \xrightarrow{\tau} P' \backslash a$

TABLE 1. Structural rules. By α–conversion the conditions marked by (*) can always be met.

2 Example.
(Client/Server Network) Consider three processes P_0, P_1, and Q. The processes P_0 and P_1 may issue requests for a certain service and the process Q issues offers for this service. The requests are labeled by r and the offers by $\bar{o}$ where $r, o \in \textit{Name}$ and $r \neq o$. The problem consists in connecting P_0 and P_1 with Q. This is to be done in a way so that P_i, $i \in \{0, 1\}$, has the possibility to claim priority for the next access to the service while a request by $P_{(i+1) \bmod 2}$ is being served.

Our solution involves four auxiliary processes: S_1, S_2, L_1, and L_2. The first two act in a semaphore–like manner while the other two form the link between P_0 or P_1, respectively, and Q. The idea is that the semaphore established by S_1 lies before the semaphore established by S_2 and that the latter one is used to enclose the access to the service offered by Q. While $P_{(i+1) \bmod 2}$ waits inside the second semaphore, then, P_i can enter the first one so as to gain priority over $P_{(i+1) \bmod 2}$ as soon as this process is released.

The individual processes and the overall system look as follows:

$S_i = Y_X[\text{get}_i \to X]$ for $i \in \{1,2\}$,
$L_0 = L_1 = Y_X[\overline{r} \to \lambda Y.\overline{\text{get}}_1 \to \lambda Y_1.\overline{\text{get}}_2 \to \lambda Y_2.(Y_1|(o \to (Y|Y_2|X)))]$, and
System $= S_1 | S_2 | (P_0|L_0)\backslash r | (P_1|L_1)\backslash r | Q$

where X, Y, Y_1, and Y_2 are arbitrary but pairwise distinct.

The reader may note the following points:

- As the auxiliary processes must not terminate the recursion context from Example 1 is employed.
- The semaphores are implemented using the rendezvous. It is crucial that the first semaphore is left (in Y_1) only after the second one has been entered.
- When a synchronization between L_i and Q takes place then Q receives residuals from P_i (in Y), S_2 (in Y_2), and L_i (in X). As soon as Q has executed the service these residuals are released simultaneously.

3 Introducing Notions of Rendezvous Based Strong Bisimulation

Notions of strong bisimulation commonly require that two related processes can match each other's initial moves so as to become related again. Since the outcome of an r- or o-transition is a triple or function, respectively, we cannot directly impose this view on our framework. An immediate solution, however, consists in requiring that triples are component-wise and functions are pointwise related. Next we formalize this intuition. At the same time we define a notion of simple rendezvous based strong bisimulation. It is much more lavish than the main one.

3 Definition. A binary relation $\mathcal{R}$ over Pr is a *(simple) rendezvous based strong simulation* if the following three statements are true for all P,Q so that $P\mathcal{R}Q$:

(i) Whenever $P \xrightarrow{a} (P', R, A)$ then, for some Q' and some S, $Q \xrightarrow{a} (Q', S, A)$ and $P'\mathcal{R}Q'$ and $R\mathcal{R}S$.

(ii) Whenever $P \xrightarrow{\overline{a}} \lambda X.T$ then, for some U, $Q \xrightarrow{\overline{a}} \lambda X.U$ and, for every R, $T[R/X]\mathcal{R}U[R/X]$ (for every R there exists an S so that $R\mathcal{R}S$ and $T[R/X]\mathcal{R}U[S/X]$).

(iii) Whenever $P \xrightarrow{\tau} P'$ then, for some Q', $Q \xrightarrow{\tau} Q'$ and $P'\mathcal{R}Q'$.

A relation $\mathcal{R}$ is a *(simple) rendezvous based strong bisimulation* if both $\mathcal{R}$ and $\mathcal{R}^{-1}$ are (simple) rendezvous based strong simulations. For all P,Q, we denote by $P \overset{r}{\sim} Q$ ($P \overset{s}{\sim} Q$) the fact that there exists a (simple) rendezvous based strong bisimulation $\mathcal{R}$ so that $P\mathcal{R}Q$.

Using standard methods (see, e.g., [Mil89]) it is straightforward to prove the following properties about (simple) rendezvous based strong bisimulation equivalence.

4 Fact.
  (1) $\overset{r}{\sim}$ ($\overset{s}{\sim}$) is the largest (simple) rendezvous based strong bisimulation.
  (2) $\overset{r}{\sim}$ ($\overset{s}{\sim}$) is an equivalence.
  (3) $P \overset{r}{\sim} Q$ ($P \overset{s}{\sim} Q$) if and only if the following six statements are true:
   (i) Whenever $P \overset{a}{\longrightarrow} (P', R, A)$ then, for some Q' and some S, $Q \overset{a}{\longrightarrow} (Q', S, A)$ and $P' \overset{r}{\sim} Q'$ and $R \overset{r}{\sim} S$ ($P' \overset{s}{\sim} Q'$ and $R \overset{s}{\sim} S$).
   (ii) Whenever $P \overset{\bar{a}}{\longrightarrow} \lambda X.T$ then, for some U, $Q \overset{\bar{a}}{\longrightarrow} \lambda X.U$ and, for every R, $T[R/X] \overset{r}{\sim} U[R/X]$ (for every R there exists an S so that $R \mathcal{R} S$ and $T[R/X] \overset{s}{\sim} U[S/X]$).
   (iii) Whenever $P \overset{\tau}{\longrightarrow} P'$ then, for some Q', $Q \overset{\tau}{\longrightarrow} Q'$ and $P' \overset{r}{\sim} Q'$ ($P' \overset{s}{\sim} Q'$).
   (iv) Whenever $Q \overset{a}{\longrightarrow} (Q', S, A)$ then, for some P' and some R, $P \overset{a}{\longrightarrow} (P', R, A)$ and $P' \overset{r}{\sim} Q'$ and $R \overset{r}{\sim} S$ ($P' \overset{s}{\sim} Q'$ and $R \overset{s}{\sim} S$).
   (v) Whenever $Q \overset{\bar{a}}{\longrightarrow} \lambda X.U$ then, for some T, $P \overset{\bar{a}}{\longrightarrow} \lambda X.T$ and, for every R, $T[R/X] \overset{r}{\sim} U[R/X]$ (for every R there exists an S so that $R \mathcal{R} S$ and $T[R/X] \overset{s}{\sim} U[S/X]$).
   (vi) Whenever $Q \overset{\tau}{\longrightarrow} Q'$ then, for some P', $P \overset{\tau}{\longrightarrow} Q'$ and $P' \overset{r}{\sim} Q'$ ($P' \overset{s}{\sim} Q'$).

5 Corollary. $P \overset{r}{\sim} Q$ implies $P \overset{s}{\sim} Q$.

Proof. Obviously, $\overset{r}{\sim}$ is a simple rendezvous based strong bisimulation. Hence, $\overset{r}{\sim} \subseteq \overset{s}{\sim}$. □

The next proposition says that rendezvous based strong bisimulation equivalence is preserved by all our operators. In its proof we can make another use of techniques developed in the context of Plain CHOCS. This is possible for two reasons: First, a synchronization in our framework is nothing but a higher order sending event where the client's residual is the process being sent. Second, here as in Plain CHOCS the handling of the restriction operator is such that action names are statically bound.

Nevertheless the language Pr can probably not meaningfully be translated to Plain CHOCS. The reason lies in the renaming operator belonging to that calculus. For this operator can be used to rename unbound action names of a process received by another process into local names of the latter one. Nothing like this is possible here. As the rendezvous requires strict static binding it would be counterintuitive.

6 Proposition. $\overset{r}{\sim}$ *is a congruence, i.e. the following five statements are true:*
  (1) If $P \overset{r}{\sim} Q$ then $a {\rightarrow} P \overset{r}{\sim} a {\rightarrow} Q$.
  (2) Let $T, U \in \text{Pr}(X)$. If, for every R, $T[R/X] \overset{r}{\sim} U[R/X]$ then $\bar{a} {\rightarrow} \lambda X.T \overset{r}{\sim} \bar{a} {\rightarrow} \lambda X.U$.
  (3) If $P \overset{r}{\sim} Q$ then $\tau {\rightarrow} P \overset{r}{\sim} \tau {\rightarrow} Q$.
  (4) If $P_1 \overset{r}{\sim} Q_1$ and $P_2 \overset{r}{\sim} Q_2$ then $P_1 + P_2 \overset{r}{\sim} Q_1 + Q_2$.

(5) If $P_1 \stackrel{r}{\sim} Q_1$ and $P_2 \stackrel{r}{\sim} Q_2$ then $P_1|P_2 \stackrel{r}{\sim} Q_1|Q_2$.
(6) If $P \stackrel{r}{\sim} Q$ then $P\backslash a \stackrel{r}{\sim} Q\backslash a$.

Proof. (Idea) By adaptation of the proof in [Tho93] that the relation $\stackrel{\cdot}{\sim}$ (applicative higher order bisimulation) over Plain CHOCS processes is a congruence. For an outline see the appendix. □

7 Conjecture. $P \stackrel{s}{\sim} Q$ implies $P \stackrel{r}{\sim} Q$.

4 An Alternative Characterization

Several authors have proposed recursive domain equations as a means to construct semantic domains for behavioural frameworks with strong bisimulation equivalence. In [Rut92], for example, the largest solution of the equation $D = \mathscr{P}(\text{Act} \times D)$ in a universe of non–well–founded sets is employed to interpret the states of standard labelled transition systems over some set Act of actions (see [Abr91] for an approach based on well–founded sets with continuous ordering). One recurring result is that two states are related by strong bisimulation equivalence if and only if their interpretations coincide.

Below we give an alternative characterization of simple rendezvous based strong bisimulation equivalence that is motivated by these works. To this end, let, for every P, the set unf(P), called the *one step unfolding* of P, be defined by

$$\text{unf}(P) = \{(a,(P',(R,A))) : P \stackrel{a}{\longrightarrow} (P',R,A)\}$$
$$\cup \{(\overline{a}, \phi_{\lambda X.T}) : P \stackrel{\overline{a}}{\longrightarrow} \lambda X.T\}$$
$$\cup \{P' : P \stackrel{\tau}{\longrightarrow} P'\}$$

where the function $\phi_{\lambda X.T} : \text{Pr} \longrightarrow \text{Pr}$ is given by $\phi_{\lambda X.T}(Q) = T[Q/X]$. Moreover let the set unf(Pr) be defined in the canonical way, i.e. unf(Pr) $= \bigcup_{P \in \text{Pr}} \text{unf}(P)$. We use this set as the basis of a logical structure that simulates the generated part of a solution of the equation

$$D = \mathscr{P}_{\text{fin}}\Big((Name \times (D \times (D \times \mathscr{P}_{\text{fin}}(Name)))) \cup (\overline{Name} \times (D \longrightarrow D)) \cup D\Big).$$

The logical structure is an instance of the concept of i–ε–structure that we define next.

8 Definition.
(1) A tuple $E = (C, \text{\textcircled{\tiny $\in$}})$ is an ε-*structure* if C is a set and \textcircled{\tiny $\in$} is a binary relation over C. The set C is the *carrier*, the elements C are the *objects*, and \textcircled{\tiny $\in$} is the ε-*relation* of E; C is usually ranged over by x, y.
(2) A triple $E = (C, \text{\textcircled{\tiny $\in$}}, I)$ is an *i-ε-structure* if $(C, \text{\textcircled{\tiny $\in$}})$ is an ε-structure, $I \subseteq C$, and \textcircled{\tiny $\in$} $\subseteq C \times (C - I)$. The elements of I are called the *individuals* of E.

In set theory ε–structures whose carriers are classes often appear in proofs of the relative consistency of set–theoretic axioms. I.e. the objects of an ε–structure play the role of sets and the ε–relation of that ε–structure plays the role of the inclusion relation between sets. Moreover, in an i–ε–structure, the individuals play the role of Urelemente. — It is merely for technical reasons that we introduce this extra component.

Our motivation for employing i–ε–structures stems from the fact that every function denotable in our formalism is indirectly self–applicable. (To see this the reader might consider processes of the form $a \to \bar{a} \to \lambda X.T | \bar{a} \to \lambda X.T$.) I.e. we have to cope with non–well–foundedness. What is crucial, i–ε–structures allow to do so by using simple relational circularity.

To rely on ε–structures for purposes outside set theory seems to be an idea first proposed by Mahr et al. in [MSU90] (see also [Mah93]). The motivation they have is to cope with self–referentiality and non–extensionality in type theory and the semantics of natural language.

Our main tool in connection with i–ε–structures is i–ε–bisimulation. This notion generalizes the notion of ε–bisimulation introduced by Aczel ([Acz88]).

9 Definition. Let $E = (C, \varepsilon, I)$ be an i–ε–structure. A binary relation $\mathcal{R}$ over $C - I$ is an *i–ε–simulation* if the following two statements are true for all $x, y \in C - I$ so that $x \mathcal{R} y$.

   (i) Whenever $x' \varepsilon x$ and $x' \in I$ then $x' \varepsilon y$.
   (ii) Whenever $x' \varepsilon x$ and $x' \in C - I$ then, for some $y' \in C - I$, $y' \varepsilon y$ and $x' \mathcal{R} y'$.

A relation $\mathcal{R}$ over $C - I$ is an *i–ε–bisimulation* of E if both $\mathcal{R}$ and $\mathcal{R}^{-1}$ are i–ε–simulations of E. For all $x, y \in C - I$, we denote by $x \overset{E}{\sim} y$ the fact that there exists an i–ε–bisimulation of E, say $\mathcal{R}$, so that $x \mathcal{R} y$.

10 Fact. Let $E = (C, \varepsilon, I)$ be an i–ε–structure.

   (1) $\overset{E}{\sim}$ is the largest i–ε–bisimulation of E.
   (2) $\overset{E}{\sim}$ is an equivalence.
   (3) For all $x, y \in C - I$, $x \overset{E}{\sim} y$ if and only if the following four statements are true:
      (i) Whenever $x' \varepsilon x$ and $x' \in I$ then $x' \varepsilon y$.
      (ii) Whenever $x' \varepsilon x$ and $x' \in C - I$ then, for some $y' \in C - I$, $y' \varepsilon y$ and $x' \overset{E}{\sim} y'$.
      (iii) Whenever $y' \varepsilon y$ and $y' \in I$ then $y' \varepsilon x$.
      (iv) Whenever $y' \varepsilon y$ and $y' \in C - I$ then, for some $x' \in C - I$, $x' \varepsilon x$ and $x' \overset{E}{\sim} y'$.

Now we can come to our i–ε–structure, called E_{Pr}, that is based on the set unf(Pr). First, a technical definition:

11 Definition. For every set χ the set $\chi\downarrow$ is defined by

$$\chi\downarrow = \{\chi' : (\exists k \geq 2)(\exists \chi_1, \ldots, \chi_k)\, \chi' = \chi_1 \in \cdots \in \chi_k = \chi\}.$$

$\chi\downarrow$ is called the *closure* of χ.

Second, we need to make clear that we use the standard set–theoretic representation of pairs and functions. I.e. a pair of the form (χ, ψ) is actually a set of the form $\{\{\chi\}, \{\chi, \psi\}\}$ and a function is a set of such sets. Third, to avoid a very technical difficulty that we do not further discuss here we regard, on the meta–level, every element of the set $Name \cup \overline{Name} \cup Pr$ as an Urelement. Having fixed all this we can finally employ the set $\text{unf}(Pr)\downarrow$ as the carrier of E_{Pr}.

12 Definition. E_{Pr} is defined by $E_{Pr} = (\text{unf}(Pr)\downarrow, \textcircled{\in}, Name \cup \overline{Name})$ where

$$\textcircled{\in} = \{(\chi, \chi') : \chi, \chi' \in \text{unf}(Pr)\downarrow \wedge \chi' \in \chi\} \cup \{(\chi, P) : P \in Pr \wedge \chi \in \text{unf}(P)\}.$$

For all $\chi, \psi \in \text{unf}(Pr)\downarrow$ we denote by $\chi \overset{e}{\sim} \psi$ the fact that $\chi \overset{E_{Pr}}{\sim} \psi$.

Judging from the operational semantics of the language Pr one might expect that $P \overset{r}{\sim} Q$ if and only if $P \overset{e}{\sim} Q$. However, it is not known whether this property holds. We have the following alternative characterization of simple rendezvous based strong bisimulation equivalence instead.

13 Theorem. $P \overset{s}{\sim} Q$ *if and only if* $P \overset{e}{\sim} Q$.

Proof. "$\Longleftarrow$": We show that the restriction of $\overset{e}{\sim}$ to $Pr \times Pr$ is a simple rendezvous based strong bisimulation.

By symmetry it suffices to show that the restriction of $\overset{e}{\sim}$ to $Pr \times Pr$ is a simple rendezvous based strong simulation (Definition 3). We just consider clause (ii) of Definition 3, the other two clauses are easier. Thus suppose $P \overset{e}{\sim} Q$ and $P \overset{\overline{a}}{\longrightarrow} \lambda X.T$. In this case, by construction of E_{Pr}, $(\overline{a}, \phi_{\lambda X.T})\textcircled{\in} P$. Hence, by repeated application of Fact 10, Part 3 there exist x,y so that $(x,y)\textcircled{\in} Q$ and $(\overline{a}, \phi_{\lambda X.T}) \overset{e}{\sim} (x,y)$. Since $\overline{a}$ is an individual, $x = \overline{a}$ and, hence, by construction of E_{Pr} there must be a U so that $y = \phi_{\lambda X.U}$ and $Q \overset{\overline{a}}{\longrightarrow} \lambda X.U$. Now suppose $(R, T[R/X])\textcircled{\in}\phi_{\lambda X.T}$. We have $\phi_{\lambda X.T} \overset{e}{\sim} \phi_{\lambda X.U}$ and, hence, again by repeated application of Fact 10, Part 3 there exist $\tilde{x}, \tilde{y}$ so that $(\tilde{x}, \tilde{y})\textcircled{\in}\phi_{\lambda X.U}$, $R \overset{e}{\sim} \tilde{x}$, and $T[R/X] \overset{e}{\sim} \tilde{y}$. Thus, by construction of E_{Pr} there must be an S so that $\tilde{x} = S$ and $\tilde{y} = U[S/X]$.

"$\Longrightarrow$": From $\overset{s}{\sim}$ we construct an equivalence $\mathcal{R}$ over the set $\text{unf}(Pr)\downarrow$ so that $\mathcal{R}$ is an i-ε-bisimulation of E_{Pr} and $\overset{s}{\sim} \subseteq \mathcal{R}$.

Let

$$\begin{aligned}
\mathcal{R}_1 &= \left\{(\{R\}, \{S\}) : R \overset{s}{\sim} S\right\}, \\
\mathcal{R}_2 &= \left\{(\{R, A\}, \{S, A\}) : R \overset{s}{\sim} S\right\}, \\
\mathcal{R}_3 &= \left\{((R, A), (S, A)) : \{R\}\,\mathcal{R}_1\,\{S\} \wedge \{R, A\}\,\mathcal{R}_2\,\{S, A\}\right\},
\end{aligned}$$

$$\mathcal{R}_4 = \Big\{(\{P'\}, \{Q'\}) : P' \overset{s}{\sim} Q'\Big\},$$

$$\mathcal{R}_5 = \Big\{(\{P', (R,A)\}, \{Q', (S,A)\}) : P' \overset{s}{\sim} Q' \wedge (R,A)\, \mathcal{R}_3\, (S,A)\Big\},$$

$$\mathcal{R}_6 = \Big\{((P', (R,A)), (Q', (S,A))) :$$
$$\{P'\}\, \mathcal{R}_4\, \{Q'\} \wedge \{P', (R,A)\}\, \mathcal{R}_5\, \{Q', (S,A)\}\Big\},$$

$$\mathcal{R}_7 = \Big\{(\{a\}, \{a\}) : \text{true}\Big\},$$

$$\mathcal{R}_8 = \Big\{(\{a, (P', (R,A))\}, \{a, (Q', (S,A))\}) : (P', (R,A))\, \mathcal{R}_6\, (Q', (S,A))\Big\},$$

$$\mathcal{R}_9 = \Big\{((a, (P', (R,A))), (a, (Q', (S,A)))) :$$
$$\{a\}\, \mathcal{R}_7\, \{a\} \wedge \{a, (P', (R,A))\}\, \mathcal{R}_8\, \{a, (Q', (S,A))\}\Big\},$$

$$\mathcal{R}_{10} = \Big\{(\{R\}, \{S\}) : R \overset{s}{\sim} S\Big\},$$

$$\mathcal{R}_{11} = \Big\{(\{R, T[R/X]\}, \{S, U[S/X]\}) :$$
$$R \overset{s}{\sim} S \wedge T, U \in \text{Pr}(X) \wedge T[R/X] \overset{s}{\sim} U[S/X]\Big\},$$

$$\mathcal{R}_{12} = \Big\{((R, T[R/X]), (S, U[S/X])) :$$
$$\{R\}\, \mathcal{R}_{10}\, \{S\} \wedge T, U \in \text{Pr}(X) \wedge \{R, T[R/X]\}\, \mathcal{R}_{11}\, \{S, U[S/X]\}\Big\},$$

$$\mathcal{R}_{13} = \Big\{(\phi_{\lambda X.T}, \phi_{\lambda X.U}) :$$
$$T, U \in \text{Pr}(X) \wedge (\forall R)(\exists S)\, (R, T[R/X])\, \mathcal{R}_{12}\, (S, U[S/X])\Big\},$$

$$\mathcal{R}_{14} = \Big\{(\{\bar{a}\}, \{\bar{a}\}) : \text{true}\Big\},$$

$$\mathcal{R}_{15} = \Big\{(\{\bar{a}, \phi_{\lambda X.T}\}, \{\bar{a}, \phi_{\lambda X.U}\}) : T, U \in \text{Pr}(X) \wedge \phi_{\lambda X.T}\, \mathcal{R}_{13}\, \phi_{\lambda X.U}\Big\},$$

$$\mathcal{R}_{16} = \Big\{((\bar{a}, \phi_{\lambda X.T}), (\bar{a}, \phi_{\lambda X.U})) :$$
$$\{\bar{a}\}\, \mathcal{R}_{14}\, \{\bar{a}\} \wedge T, U \in \text{Pr}(X) \wedge \{\bar{a}, \phi_{\lambda X.T}\}\, \mathcal{R}_{15}\, \{\bar{a}, \phi_{\lambda X.U}\}\Big\},$$

$$\mathcal{R}_{17} = \Big\{(P', Q') : P' \overset{s}{\sim} Q'\Big\}, \text{ and}$$

$$\mathcal{R} = \text{Id}_{\mathscr{P}_{\text{fin}}(\text{Act})} \cup \overset{s}{\sim} \cup\, \mathcal{R}_1 \cup \cdots \cup \mathcal{R}_{17}.$$

For the sake of better presentational compatibility with the preceding material the defining equations for $\mathcal{R}_1$–$\mathcal{R}_{17}$ contain the following redundancies:

$$\mathcal{R}_1 = \mathcal{R}_4 = \mathcal{R}_{10} \subsetneq \mathcal{R}_{11} \text{ and } \mathcal{R}_{17} = \overset{s}{\sim}.$$

Moreover let us make clear that

$$\mathcal{R}_{11} = \{(\{P_1, P_2\}, \{Q_1, Q_2\}) : P_1 \overset{s}{\sim} Q_1 \wedge P_2 \overset{s}{\sim} Q_2\}.$$

As it is not hard to see $\mathcal{R}$ is indeed an equivalence relation. Revisiting the defining equations for $\mathcal{R}_1$–$\mathcal{R}_{17}$ it is not hard to see too that $\mathcal{R}$ is an i–ε–simulation as far as elements of the set $\text{unf}(\text{Pr}){\downarrow} - \text{Pr}$ are concerned. Otherwise suppose $P \mathcal{R} Q$ and $x \textcircled{\in} P$. By construction of E_{Pr} the object x must be of the form $(a, (P', (R, A)))$, $(\bar{a}, \phi_{\lambda X.T})$, or P'. We treat only the second case, the other two are easier. In this case, $P \xrightarrow{\overline{a}} \lambda X.T$. $P \mathcal{R} Q$ is synonymous with $P \overset{s}{\sim} Q$ and, hence, there must be a U so that $Q \xrightarrow{\overline{a}} \lambda X.U$ and for every R there exists an S so that $R \overset{s}{\sim} S$ and $T[R/X] \overset{s}{\sim} U[S/X]$. Thus, $(\overline{a}, \phi_{\lambda X.U}) \mathbin{\text{\textcircled{$\in$}}} Q$ and $(\overline{a}, \phi_{\lambda X.T}) \mathcal{R}_{13} (\overline{a}, \phi_{\lambda X.U})$, i.e. $(\overline{a}, \phi_{\lambda X.T}) \mathcal{R} (\overline{a}, \phi_{\lambda X.U})$.

By symmetry we can conclude that $\mathcal{R}$ is an i–ε–bisimulation. This completes the proof of Theorem 13. □

5 Conclusion

In Section 3 and Section 4 we have introduced two notions of rendezvous based strong bisimulation. We have proven that the equivalence emerging from the first one is a congruence, and we have given an alternative characterization of the equivalence emerging from the second one in terms of i–ε–bisimulation. Unfortunately there is not yet any similar characterization of the first equivalence. Conversely, it is not known whether the second equivalence is a congruence.

These problems would come to a very nice solution if we could prove that $P \overset{r}{\sim} Q$ if and only if $P \overset{s}{\sim} Q$. Corollary 5 covers the forward direction, so the question actually is How could we prove Conjecture 7? To this end, let us observe that all a server can do with a client residual is to start arbitrary many copies of it at different points of time. Thus we may state the following additional conjecture.

14 Conjecture. *Let, for some X, $T, U \in \mathrm{Pr}(X)$. If, for any $a \in \mathit{Name} - (\mathrm{fn}(T) \cup \mathrm{fn}(U))$, $T[a \to \mathrm{nil}/X] \overset{r}{\sim} U[a \to \mathrm{nil}/X]$ then, for every P, $T[P/X] \overset{r}{\sim} U[P/X]$.*

Proof. (Rough idea) Suppose $T[a \to \mathrm{nil}/X] \overset{r}{\sim} U[a \to \mathrm{nil}/X]$ for some $a \in \mathit{Name} - (\mathrm{fn}(T) \cup \mathrm{fn}(U))$. R–transitions labelled by a, then, mark the starting points for client residuals in $T[a \to \mathrm{nil}/X]$ and $U[a \to \mathrm{nil}/X]$. These starting points are the same in $T[P/X]$ and $U[P/X]$ when P is arbitrary. Hence, $T[P/X] \overset{r}{\sim} U[P/X]$. □

In future work the author wants to give a full proof of this property. It seems to be the crucial step on the way of proving that Conjecture 7 does indeed hold.

References

[Abr91] S. Abramsky. A Domain Equation for Bisimulation. *Information and Computation*, (92):161–218, 1991.
[Acz88] P. Aczel. *Non–well–founded Sets*. Stanford University, 1988.
[Bak89] J.W. de Bakker. Designing Concurrency Semantics. Technical Report CS-R8902, Centre for Mathematics and Computer Science, Amsterdam, 1989.
[Bal94] M. Baldamus. Tau–less CCS with a Remote Procedure Call Primitive. Technical Report 1994/6, Fachbereich Informatik, Technische Universität Berlin, 1994.
[BV91a] J.W. de Bakker and E.P. de Vink. CCS for OO and LP. In S. Abramsky and T.S.E. Maibaum, editors, *Theory and Practice of Software Development, Advances in Distributed Computing and Combining Paradigms for Software Development*, LNCS 494, pages 1–28. Springer, 1991.

[BV91b] J.W. de Bakker and E.P. de Vink. Rendez–Vous with Metric Semantics. In E.H. Aarts, J. van Leeuwen, and M. Rem, editors, *Parallel Languages and Architectures Europe. Volume II: Parallel Languages*, LNCS 506, pages 27–57. Springer, 1991.

[BZ83] J.W. de Bakker and J.I. Zucker. Processes and a Fair Semantics for the Ada Rendez–Vous. In J. Diaz, editor, *Automata, Languages and Programming*, LNCS 154, pages 52–66. Springer, 1983.

[DoD83] United States Department of Defense. *Reference Manual for the Ada Programming Language*, 1983. ANSI/MIL–STD–1815A–1983.

[Egg91] G. Egger. *Integration applikativer und prozeßorientierter Programmierung*. PhD thesis, Fachbereich Informatik, Technische Universität Berlin, 1991. In German.

[Fra94a] T. Frauenstein. Eine dienstorientierte applikative Prozeßsprache und ihre Fundierung durch den π–Kalkül. Technical Report 1994/1, Fachbereich Informatik, Technische Universität Berlin, 1994. In German.

[Fra94b] T. Frauenstein. Remote Function Call within an Applicative Higher–Order Process Calculus. Technical Report 1994/27, Fachbereich Informatik, Technische Universität Berlin, 1994.

[HL93] M. Hennessy and H. Lin. Proof Systems for Message–passing Process Algebras. In E. Best, editor, *Concur '93*, LNCS 715. Springer, 1993.

[Mah93] B. Mahr. Applications of Type Theory. In M.-C. Gaudel and J.-P. Jouannaud, editors, *Theory and Practice of Software Development*, LNCS 668, pages 343–355. Springer, 1993.

[Mil89] R. Milner. *Communication and Concurrency*. Prentice Hall International, 1989.

[MP89] R. Milner and J. Parrow. A Calculus of Mobile Processes, I and A Calculus of Mobile Processes, II. Technical Reports ECS–LFCS–89–85 and –86, Laboratory for Foundations of Computer Science, University of Edinburgh, 1989.

[MSU90] B. Mahr, W. Sträter, and C. Umbach. Fundamentals of a Theory of Types and Declarations. Technical Report KIT 82, Fachbereich Informatik, Technische Universität Berlin, 1990.

[Rut92] J.J.M.M. Rutten. Processes as Terms: Non–well–founded Models for Bisimulation. *Mathematical Structures in Computer Science*, 2:257–275, 1992.

[Tho93] B. Thomsen. Plain CHOCS — A Second Generation Calculus for Higher Order Processes. *Acta Informatica*, 30(1):1–59, 1993.

Appendix

Below we give an outline of the proof of Proposition 6. To repeat what was stated above about this proof, it is an adaptation of the proof in [Tho93] that the relation $\overset{r}{\sim}$ over Plain CHOCS processes is a congruence. As the framework presented here is (intentionally) not truly higher order several modifications and simplifications arise.

Statement. $\overset{r}{\sim}$ is a congruence, i.e. the following five statements are true:

(1) If $P \overset{r}{\sim} Q$ then $a \to P \overset{r}{\sim} a \to Q$.
(2) Let $T, U \in \text{Pr}(X)$. If, for every R, $T[R/X] \overset{r}{\sim} U[R/X]$ then $\bar{a} \to \lambda X.T \overset{r}{\sim} \bar{a} \to \lambda X.U$.
(3) If $P \overset{r}{\sim} Q$ then $\tau \to P \overset{r}{\sim} \tau \to Q$.
(4) If $P_1 \overset{r}{\sim} Q_1$ and $P_2 \overset{r}{\sim} Q_2$ then $P_1 + P_2 \overset{r}{\sim} Q_1 + Q_2$.
(5) If $P_1 \overset{r}{\sim} Q_1$ and $P_2 \overset{r}{\sim} Q_2$ then $P_1|P_2 \overset{r}{\sim} Q_1|Q_2$.
(6) If $P \overset{r}{\sim} Q$ then $P\backslash a \overset{r}{\sim} Q\backslash a$.

Proof. (Outline)

As to (2): By Fact 4.

As to (1) and (3)–(6): Here it is useful to introduce the following notational conventions:

- By $\bar{P}, \bar{Q}, \ldots$ we denote finite vectors of variables and by $\bar{X}, \bar{Y}, \ldots$ we denote finite ordered sets of variables.
- For all $\bar{P}, \bar{Q}$ of equal length we denote by $\bar{P} \stackrel{r}{\sim} \bar{Q}$ the fact that $\bar{P}$ and $\bar{Q}$ are component–wise related via $\stackrel{r}{\sim}$.
- For all $T, \bar{P}, \bar{X}$ so that, for some k, $\bar{P} = (P_1, \ldots, P_k)$ and $\bar{X} = \{X_1, \ldots, X_k\}$ we denote by $T[\bar{P}/\bar{X}]$ the term $T[P_1/X_1]\ldots[P_k/X_k]$.

It will always be clear from the context which vectors and/or ordered sets of variables must have equal size. Thus, from now on we do not make this side condition explicit anymore.

To start with the actual proof, let

$$\mathcal{R} = \{(T[\bar{P}/\bar{X}], T[\bar{Q}/\bar{X}]) : \text{fv}(T) \subseteq \bar{X} \text{ and } \bar{P} \stackrel{r}{\sim} \bar{Q}\}$$

and let $\mathcal{C}$ be the transitive closure of $\mathcal{R}$. We show that $\mathcal{C}$ is a rendezvous based strong bisimulation. (1) and (3)–(6) then follow because $\mathcal{C}$ contains every pair that is based on a context of the form $a \rightarrow X$, $\tau \rightarrow X$, $X_1 + X_2$, $X_1 | X_2$, or $X \backslash a$.

First we show that the following three statements are true for all $T, \bar{X}, \bar{P}, \bar{Q}$ so that $\text{fv}(T) \subseteq \bar{X}$ and $\bar{P} \stackrel{r}{\sim} \bar{Q}$:

(i) Whenever $T[\bar{P}/\bar{X}] \stackrel{a}{\longrightarrow} (P', R, A)$ then, for some Q' and some S, $T[\bar{Q}/\bar{X}] \stackrel{a}{\longrightarrow} (Q', S, A)$ and $P' \mathcal{C} Q'$ and $R \mathcal{C} S$.

(ii) Whenever $T[\bar{P}/\bar{X}] \stackrel{\bar{a}}{\longrightarrow} \lambda Y.U$ then, for some V, $T[\bar{Q}/\bar{X}] \stackrel{\bar{a}}{\longrightarrow} \lambda Y.V$ and, for every R, $U[R/Y] \mathcal{C} V[R/Y]$.

(iii) Whenever $T[\bar{P}/\bar{X}] \stackrel{\tau}{\longrightarrow} P'$ then, for some Q', $T[\bar{Q}/\bar{X}] \stackrel{\tau}{\longrightarrow} Q'$ and $P' \mathcal{C} Q'$.

As in [Tho93], the proof is done by induction on the length of the inference used to establish any particular transition of $T[\bar{P}/\bar{X}]$ and by case distinction on the structure of T. Many subcases arise. However, all those are alike where, for some Y, $T = Y$. For, under these circumstances, among $\bar{P}$ and $\bar{Q}$ there must be P, Q so that $T[\bar{P}/\bar{X}] = P$, $T[\bar{Q}/\bar{X}] = Q$, and $P \stackrel{r}{\sim} Q$; this makes it straightforward to prove the desired properties using Fact 4 and the fact that $\stackrel{r}{\sim} \subseteq \mathcal{R} \subseteq \mathcal{C}$.

From the set of remaining subcases we only present the most complex one; it occurs if there exist T_1, T_2 so that $T = T_1|T_2$, $T[\bar{P}/\bar{X}] \stackrel{\tau}{\longrightarrow} (R|U[P'/Y])\backslash A$, and, by shorter inferences, $T_1[\bar{P}/\bar{X}] \stackrel{a}{\longrightarrow} (P', R, A)$ and $T_2[\bar{P}/\bar{X}] \stackrel{\bar{a}}{\longrightarrow} \lambda Y.U$. Then, $T[\bar{Q}/\bar{X}] = T_1[\bar{Q}/\bar{X}]|T_2[\bar{Q}/\bar{X}]$ and, hence, by induction there exist Q', S, V so that $T_1[\bar{Q}/\bar{X}] \stackrel{a}{\longrightarrow} (Q', S, A)$, $P' \mathcal{C} Q'$, $R \mathcal{C} S$, $T_2[\bar{Q}/\bar{X}] \stackrel{\bar{a}}{\longrightarrow} \lambda Y.V$ and, for every K, $U[K/Y] \mathcal{C} V[K/Y]$. Thus, by the rule for synchronization, $T[\bar{Q}/\bar{X}] \stackrel{\tau}{\longrightarrow}$

$(S|V[Q'/Y])\backslash A$. To show

(*) $\qquad (R|U[P'/Y])\backslash A \, \mathcal{C} \, (S|V[Q'/Y])\backslash A$

we employ the following four technical lemmas:

Lemma. *Let $T \in \mathrm{Pr}(X)$. For all P, Q, if $P \mathcal{R} Q$ then $T[P/X] \mathcal{R} T[Q/X]$.*

Lemma. *Let $T \in \mathrm{Pr}(X)$. For all P, Q, if $P \mathcal{C} Q$ then $T[P/X] \mathcal{C} T[Q/X]$.*

Lemma. *For all P_1, P_2, Q_1, Q_2, if $P_1 \mathcal{R} Q_1$ and $P_2 \mathcal{R} Q_2$ then $P_1|P_2 \mathcal{R} Q_1|Q_2$.*

Lemma. *For all P_1, P_2, Q_1, Q_2, if $P_1 \mathcal{C} Q_1$ and $P_2 \mathcal{C} Q_2$ then $P_1|P_2 \mathcal{C} Q_1|Q_2$.*

Their proofs can easily be derived from the proofs of some corresponding lemmas in [Tho93]; in that, the first and the third are actually needed to prove the second or fourth, respectively.

To return to (*), obviously we have $U[P'/Y] \, \mathcal{C} \, V[P'/Y]$. Moreover, by the second lemma, $V[P'/Y] \, \mathcal{C} \, V[Q'/Y]$. Hence, by transitivity, $R \, \mathcal{C} \, S$, and the fourth lemma, $R|U[P'/Y] \, \mathcal{C} \, S|V[Q'/Y]$. Another application of the second lemma leads to the desired conclusion.

To return to the main proposition, let us consider some sequence of the form

$$T_1[\bar{P}_1/\bar{X}_1] \, \mathcal{R} \, T_1[\bar{Q}_1/\bar{X}_1] = T_2[\bar{P}_2/\bar{X}_2]$$

$$\ldots$$

$$T_{k-1}[\bar{Q}_{k-1}/\bar{X}_{k-1}] = T_k[\bar{P}_k/\bar{X}_k] \, \mathcal{R} \, T_k[\bar{Q}_k/\bar{X}_k].$$

Using the property above a simple induction on k suffices to establish that $T_k[\bar{Q}_k/\bar{X}_k]$ can simulate $T_1[\bar{P}_1/\bar{X}_1]$ as required.

Finally let us observe that $\mathcal{C}$ is symmetric because $\overset{r}{\sim}$ and $\mathcal{R}$ are symmetric. Hence, $\mathcal{C}$ is a rendezvous based strong bisimulation. This completes the outline of the proof of Proposition 6. □

Transient Analysis of Real-Time Systems Using Deterministic and Stochastic Petri Nets

Varsha Mainkar[1] and Kishor S. Trivedi[2]

[1] Dept. of Computer Science, Duke University, Durham, NC 27708, USA
[2] Dept. of Electrical Engineering, Duke University, Durham, NC 27708, USA

Abstract. We present an analysis of a real-time system which has fixed-priority scheduling of aperiodic and periodic tasks. We assume that the cycle times of periodic tasks are multiples of each other, task execution times have phase-type distributions and aperiodic tasks arrive from Poisson sources. Since interarrival times of periodic tasks are constant, deterministic and stochastic Petri nets (DSPN) are an appropriate model. Using Markov regenerative theory to solve the DSPN model, the time-dependent behavior of the real-time system is studied. The steady-state behavior of such a system is shown to be periodic, i.e., there is no limiting probability distribution. Further, the *response time distribution* of an aperiodic task is also computed numerically. To our knowledge, this is the first time an analytical evaluation of a real-time system has been carried out at this level of detail.

1 Introduction

Performance evaluation of real-time systems offers unique challenges to the modeling techniques of today. The timing requirements of real-time systems are very stringent, while the software and hardware structure can be quite complex. Some of the features that characterize real-time systems are (1) a software task structure that includes periodic as well as aperiodic tasks, (2) *constant time* events such as cycle times of periodic tasks, (3) tasks that must meet deadlines. In spite of all these characteristics, existing literature shows a continued use of simple queueing models such as the M/M/1 queue [7, 8, 9] for computation of performance measures. Such estimates serve only as rough guidelines for design decisions. It is quite clear that M/M/1 queues do not adequately represent a real-time system because there are multiple sources of aperiodic and periodic task arrival. Continuous-time Markov chain (CTMC) [13] models *cannot* represent a real-time system because of the deterministic nature of some events. Discrete-time Markov chains (DTMC) [13] may be applied, however, we must then observe the system at only specific epochs in time, and this means loss of information.

Furthermore, there is not much evidence of work in computation of more "advanced" measures such as probabilities of missed deadlines. Traditional analyses of real-time systems concentrate on mean value analysis and steady-state behavior. Literature shows use of average performance measures or best/worst case performance measures [9]. However, time-dependent behavior of a system is important in cases when a system does not exhibit a limiting behavior or for studying the behavior of the system under transient overloads. The response time distribution of a task is also an important measure, which tells us the probability of a task missing its deadline. This distribution can be computed only by transient analysis of a system model. Such computation can answer what-if questions such as how many periodic tasks can be run and how frequently, and what load of aperiodic tasks can be sustained so as to be able to meet deadlines with high probability.

It is quite apparent then, that there is a need for better models of real-time systems. Some work related to formal design and analysis of real-time systems has been reported in [1]. Response time distribution of a task in a real-time system was also evaluated in [12], using stochastic reward nets, which are solved by generation of the underlying CTMC. However, there is much scope for improvement. We require two important advancements: better models, and more useful measures using these models. For the former case, deterministic and stochastic Petri nets may be used, since they have the capability of modeling deterministic event times. For the latter, recent developments in transient analysis of DSPNs offer a great opportunity in studying time-dependent behavior of systems, and computing measures such as response time distribution. We have shown in this paper that using more advanced models such as DSPN gives significantly different results than the ones resulting from standard techniques using Markov modeling.

The rest of the paper is as follows: in Section 2 we provide a brief introduction to DSPNs. In Section 3 we introduce the basic characteristics of real-time systems and describe in detail the features of the real-time system that we consider. In Section 4 we present its DSPN model. In Section 5 we describe how the transient solution of this DSPN can be obtained. In Section 6 we present numerical results. In Section 7 we present the DSPN model used to compute the response time distribution of an aperiodic task. We conclude the paper in Section 8.

2 Overview of DSPNs

A Petri net [11] is a directed bipartite graph with two types of nodes called **places** (represented by circles) and **transitions** (represented by rectangles or bars). Directed arcs (represented by arrows) connect places to transitions, and vice versa. If an arc exists from a place (transition) to a transition (place), then the place is called an input (output) place of that transition. Places may contain **tokens** (represented by dots). The state of a Petri net is defined by a vector of the number of tokens in each place, called a **marking** of the Petri net. The notation $\#(p, m)$ is used to denote the number of tokens in a place p in a marking m. When the context is clear, the "m" is dropped from the notation. Typically, places represent conditions or resources, and transitions represent events or choices. Tokens may flow along the directed arcs of the Petri net according to some rules, thus reflecting the changes in the system due to various events.

A **multiplicity** is a non-negative integer that may be associated with an input or output arc. A transition is said to be **enabled** if each of its input places contains at least as many tokens as that input arc's multiplicity. An enabled transition can **fire**. When it fires, as many tokens as an input arc's multiplicity are removed from the corresponding input place, and as many tokens as an output arc's multiplicity are deposited in the corresponding output place. A set of transitions is said to be **conflicting** when the firing of one disables the rest. Transitions may be assigned priorities that can be used to resolve conflicts between transitions.

Structural extensions to Petri nets include **inhibitor** arcs (denoted by an arc with a circle instead of an arrow head), which connect places to transitions. A transition can be enabled only if the number of tokens in its inhibitor place is less than the multiplicity of the inhibitor arc.

Timed Petri nets associate a time delay with transitions. The **generalized stochastic Petri net** (GSPN) is a type of Petri net that allows transitions to have an exponentially distributed time delay (**timed** transitions, represented by rectangles) or a zero time delay (**immediate** transitions, represented by bars). A marking of a GSPN is said to be **vanishing** if at least one immediate transition is enabled in it and is said to be **tangible** otherwise. Conflicts among immediate transitions in a vanishing marking may be resolved by assigning probabilities to conflicting sets of immediate transitions. A GSPN can be solved analytically by generating its underlying stochastic process (termed the **marking process**) which is a CTMC. This CTMC is the one generated by the tangible markings of the GSPN. Several extensions have been proposed under the formalism of **stochastic reward nets** which keep the underlying stochastic process of the Petri net a CTMC, but make specification easier [4]: The firing rate of the timed transitions may be marking-dependent. The probability of firing of an immediate transition may also be marking-dependent. Multiplicities of arcs may be marking-dependent. Such arcs are termed **variable cardinality** arcs. Further, enabling functions or **guards** may be associated with transitions. Guards are marking-dependent predicates which must be satisfied for transitions to be considered enabled.

Deterministic and stochastic Petri nets are timed Petri nets that allow immediate transitions and timed transitions, where the timed transitions may have either exponentially distributed firing time (termed EXP transitions) or deterministic firing time (termed DET transitions). The condition is that at any time *at most one deterministic transition may be enabled*. When this condition is met, it can be shown that the stochastic process corresponding to the tangible markings is the **Markov regenerative process** [3] (also called semi-regenerative process in [2]). A Markov regenerative process can be described informally as a process in which we can find an embedded sequence of time points at which the past history of the evolution of the process may be "forgotten". The future evolution of the process depends only on the current state at these embedded time points. Note that between these **Markov regeneration epochs** the state of the process may change but this change does not imply a Markov regeneration. This is more general than the CTMC where at *any* time the past history is summarized in its current state, and the semi-Markov chain, where each state-change implies regeneration (in the Markovian sense).

For a formal definition of an MRGP we first need to define a Markov renewal sequence:

Definition 1. [2, 6]
A sequence of bivariate random variables $\{(Y_n, T_n), n \geq 0\}$ is called a **Markov renewal sequence** if

1. $T_0 = 0$, $\forall n > 0$, $T_{n+1} > T_n$ and $Y_n \in \Omega \subset \mathbb{Z}$, the set of integers
2. $\forall i, j \in \Omega$,

$$P\{Y_{n+1} = j,\ T_{n+1} - T_n \leq t \mid Y_n = i, T_n, Y_{n-1}, T_{n-1}, ..., Y_0, T_0\}$$
$$= P\{Y_{n+1} = j,\ T_{n+1} - T_n \leq t \mid Y_n = i, T_n\} \quad \text{(Markov Dependence)}$$
$$= P\{Y_1 = j,\ T_1 \leq t \mid Y_0 = i\} \quad \text{(Time Homogeneity)}. \qquad (1)$$

The MRGP is defined as follows:

Definition 2. [2, 6]
A stochastic process $\{Z(t), t \geq 0\}$ is called a **Markov regenerative process** (or a semi-regenerative process), if there exists a Markov renewal sequence $\{(Y_n, T_n), n \geq 0\}$ of random variables such that all the conditional finite dimensional distributions of $\{Z(T_n + t), t \geq 0\}$ given $\{Z(u), 0 \leq u \leq T_n, Y_n = i\}$ are the same as those of $\{Z(t), t \geq 0\}$ given $Y_0 = i$.

This definition implies that

$$P\{Z(T_n + t) = j \mid Z(u), 0 \leq u \leq T_n, Y_n = i\} = P\{Z(t) = j \mid Y_0 = i\}. \qquad (2)$$

This means that the Markov regenerative process $\{Z(t), t \geq 0\}$ does not have the Markov property in general, but there is a sequence of embedded time points $(T_0, T_1, ..., T_n, ...)$ such that the states $(Y_0, Y_1, ..., Y_n, ...)$, respectively of the process at these time points satisfy the Markov property. These time points are the Markov regeneration epochs. At time T_n, it does not matter what states the process $\{Z(t), t \geq 0\}$ has visited until reaching Y_n at time T_n. The state $Z(T_n)$ is the only needed information for the future evolution of $Z(T_n + t)$, $t \geq 0$. Note that the stochastic process $\{Y_n, n = 0, 1, \ldots\}$ is a discrete time Markov chain. Further, let $N(t) = \sup\{n \geq 0 : T_n \leq t\}$. Then the stochastic process $\{X(t), t \geq 0\}$ where $X(t) = Y_{N(t)}$ is a semi-Markov process.

The following conditional probabilities are necessary to be defined for the analysis of an MRGP:

$$K_{ij}(t) = P\{Y_1 = j,\ T_1 \leq t \mid Y_0 = i\} \qquad i, j \in \Omega. \qquad (3)$$

and
$$E_{ij}(t) = P\{Z(t) = j, T_1 > t \mid Y_0 = i\}. \tag{4}$$

The matrix $K = [K_{ij}(t)]$ is termed the **global kernel** [3] and is the joint conditional probability of the time to the next Markov regeneration and the state right after the next Markov regeneration given the state at the current Markov regeneration. The matrix $E_{ij}(t)$ (called the **local kernel**) describes the evolution of the MRGP between two Markov regeneration epochs. The matrices K and E will be used in computing the transient probability :

$$P_{ij}(t) = P\{Z(t) = j \mid Z(0) = Y_0 = i\} \tag{5}$$

Let $K*P$ denote a matrix whose (i,j)th element is $\sum_u K_{iu} * P_{uj}(t)$ where

$$K_{iu} * P_{uj}(t) = \int_0^t dK_{iu}(x) P_{uj}(t-x).$$

Then the transient probability $[P_{ij}(t)]$ satisfies the Markov renewal equation [2]:

$$P(t) = E(t) + K*P(t). \tag{6}$$

where $P = P_{ij}(t)$.

It was shown in [3] that if $\{Z(t), t \geq 0\}$ is the stochastic process corresponding to the tangible markings of the DSPN, then $Z(t)$ is an MRGP. The Markov regeneration epochs T_i were defined as

- the time of the next state change, if no deterministic transition is enabled in the current state.
- the time of the firing of the deterministic transition, if one deterministic transition is enabled in the current state.

The matrices $K(t)$ and $E(t)$ corresponding to these Markov regeneration epochs were defined for any general DSPN in [3]. In this paper we shall not present those general equations; rather, we shall present equations specialized to the DSPN model of the real-time system.

3 System Description

3.1 Real Time System Characteristics

A design of a real-time system is significantly different from a normal computer system. A real-time system must be designed so that its behavior is *predictable*. This is because of the strict timing constraints that a real-time system must satisfy. The components of a real-time system relevant to performance analysis are a processor, a set of periodic and aperiodic tasks and a scheduler. Each periodic task i, arrives to the system every c_i time units. Periodic tasks typically perform some routine monitoring of the system. Arrival of aperiodic tasks is usually caused by external sources such as changes in the environment. The scheduler is an important component of a realtime system. The scheduler is responsible to choose the next task to be executed. The scheduling discipline is usually *priority*, as this can ensure predictable response times for the higher priority tasks. Tasks may have deadlines. Deadlines of tasks are classified as *hard* or *soft*. Missing a hard deadline can have catastrophic effects, whereas missing a soft deadline will only cause minor performance degradation. Task execution times may vary a little, and are hence stochastic.

For us to evaluate this system analytically, we must make some assumptions. These are listed next.

3.2 Assumptions

The real-time system considered in this paper has the following features:

- There are m periodic tasks, $1, 2, \ldots, m$. Each periodic task k has a cycle time of $c_k \tau$, and a priority p_k. The execution time of the task can be *phase-type*, however, assume for simplicity that it is exponentially distributed with mean $1/\mu_k$.
- The timer for each periodic task begins at the same time.
- There are n aperiodic task sources, $m+1, m+2, \ldots, m+n$. An aperiodic task of type $m+k$ arrives to the real-time system from a Poisson source at the rate λ_k. The execution time of the task can be phase-type, however, assume for simplicity that it is exponentially distributed with mean $1/\mu_{m+k}$. The aperiodic task also has an assigned priority of p_{m+k}.
- Each task has its own queue.
- We assume the real-time system to be soft.
- There is a limit on the number of instances of a particular task that can be waiting for execution at any time. Let that limit be denoted by m_k for task k. Periodic tasks are not scheduled when this limit is reached, and if an arriving aperiodic task finds that the limit for its own type has been reached, it is lost.
- There is a single processor.
- Scheduling is static and fixed priority. Whenever a task completes, the next instance of a task with the highest priority (including periodic and aperiodic tasks) is executed. The scheduling is also *non-preemptive*.
- Processing time required by the scheduler to choose the next task for execution is negligible.

4 The DSPN Model of the system

In this and the next section, we present the DSPN model of the real-time system, and present equations for transient analysis that are based on the theory developed in [3] but that have been simplified to a great extent because of the special nature of our problem.

With the assumptions outlined in the earlier section, a DSPN that models the state of the system can be built. Figure 1 shows a DSPN for a system with two periodic tasks and one aperiodic task. In this DSPN, we have used many of the *structural* extensions defined for stochastic reward nets (such as marking dependent multiplicities, marking dependent firing rate for transitions with exponentially distributed firing time) which make our task of specification easier, while still maintaining the analytical tractability of the DSPN.

In the figure, a token in place P_{avail} models an idle processor. Transition *timer* represents a clock tick which is used to schedule the periodic tasks. The time, τ, of this transition is equal to the greatest common divisor of the cycle times of the periodic tasks. Places P_1 and P_2 represent periodic tasks 1 and 2. When c_1 tokens accumulate in place P_1, the timer has fired c_1 times, and thus an instance of task 1 is created and is waiting to be scheduled. Similarly c_2 tokens accumulated in place P_2 denotes the arrival of periodic task 2. The cycle time of a periodic task i is enforced by setting the multiplicity of the arc from P_i to immediate transition t_i equal to c_i. Thus transition t_i is enabled only if there are at least c_i tokens in P_i. Transition t_{arrive} and place P_3 represent the arrival of an aperiodic task. Transition t_{arrive} has a firing rate λ_3. Transitions *timer* and t_{arrive} are always enabled. Further, transition t_i has priority p_i which models the static fixed priority scheduling policy.

When transition t_i fires, it puts i tokens in the place P_{exec}. This keeps track of the task that is executing in the processor. The rate of transition T_{exec} is marking-dependent and is equal to μ_k when there are k tokens in place P_{exec}. The arc from P_{exec} to T_{exec} is a variable cardinality arc, whose multiplicity is given by $\#(P_{exec})$ when $\#(P_{exec}) > 0$ and 1 otherwise. Thus when T_{exec} fires it, removes all the tokens from P_{exec}.

The transitions f_i model the limit on the number of waiting tasks in the system. Thus any arrival which occurs when the limit has been reached are "rejected". This is modeled by setting the multiplicity of the arc from P_i to f_i equal to c_i ($c_3 = 1$), and a guard for transition f_i which is true when $\#(P_i) = (m_i + 1)c_i$ and false otherwise. Thus if the buffer limit of task 2 is 2 and the period is 2τ, 2 tokens will be immediately "flushed out" when the number of tokens in P_2 reaches 6. This models the rejection of the task. The priority of transitions f_i is higher than all the other transitions in the net.

If tasks have phase-type execution time, it is very easy to extend the DSPN in Figure 1 to represent phase-type distributions. The transition T_{exec} will have to be replaced by a subnet for each task representing the phase-type execution time.

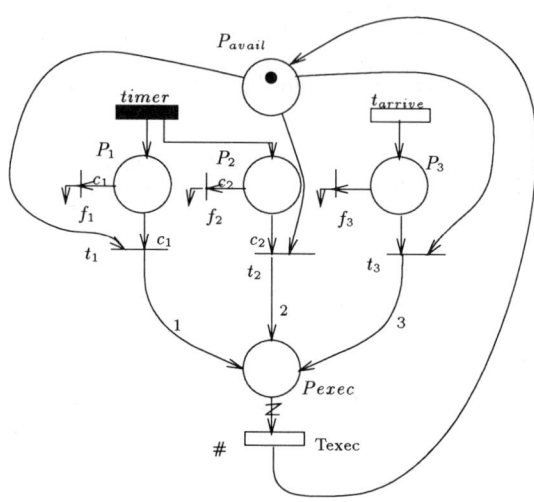

Fig. 1. DSPN model of the real-time system

5 Transient Solution of the DSPN Model

The tangible markings of The DSPN of Figure 1 define a stochastic process. Since the DSPN satisfies the condition of at most one DET transition being enabled at any time, this marking process is the Markov regenerative process $Z(t)$. The transient solution of this MRGP is simplified by the fact that there is one deterministic transition (the *timer*) enabled all the time which is not competitive with any other transition. Thus the Markov regeneration epochs of $Z(t)$ are the times of the firing of the transition *timer*, i.e., $0, \tau, 2\tau, \ldots$. Between two firings of the *timer*, the two other events that may bring about a state change are arrival of an aperiodic task and completion of a task on the processor. The timings of these events are exponentially distributed and may be described by a CTMC generator matrix. In particular, let Q denote the infinitesimal generator matrix corresponding to all the transitions between tangible markings caused by the firing of EXP transitions. For instance, consider the tangible marking $M_1 = (\#(P_{avail}), \#(P_1), \#(P_2), \#(P_3), \#(P_{exec})) = (1,0,0,0,0)$. This tangible marking directly reaches tangible marking $M_2 = (0,0,0,0,3)$ by the firing of transition t_{arrive} which has exponentially distributed firing time. Therefore, $Q_{12} = \lambda_3$.

The state changes due to the firing of the deterministic transition $timer$ are recorded in a matrix Δ. Then, Δ is defined as:

$$\Delta_{ij} = \Pr\{\text{next tangible marking is } M_j \mid \text{current tangible marking is } M_i \text{ and } timer \text{ fires}\}.$$

For instance, suppose $c_1 = 1$ and $c_2 = 2$; then the tangible marking reached from marking M_1 defined above, due to firing of transition $timer$ is $M_3 = (0,1,0,0,1)$, with probability one. Therefore, $\Delta_{13} = 1$.

Now, suppose we are given the initial probability vector, $p(0)$ of the system. Suppose we need to find the probability that the system is in state j at time t, where $t < \tau$. Obviously, this probability is simply $p_j(t)$, where $p(t) = [p_j(t)]$ is given by,

$$p(t) = p(0)e^{Qt}. \qquad (7)$$

This is because the only state changes that can occur within the interval $[0, \tau)$ are the ones caused by EXP transitions.

The state probability vector at time $t = \tau + s$, where $0 \le s < \tau$, can be derived as follows: the transient probability vector of the system just before the timer first fires is given, according to Equation (7), by $p(\tau^-) = p(0)e^{Q\tau}$. The probability vector at the time the timer fires, is given by $p(\tau) = p(0)e^{Q\tau}\Delta$. Next, the evolution over the subsequent time period $(\tau, \tau+s)$ is again due to only exponential transitions and will be described by the Q matrix. Finally, the transient probability vector at time t is given by :

$$p(t) = p(\tau + s) = p(0)e^{Q\tau}\Delta e^{Qs}. \qquad (8)$$

In the notation described in Section 2, the matrix $E(t)$ corresponds to e^{Qt} for $t < \tau$ and zero otherwise, while the matrix $K(t)$ corresponds to $e^{Q\tau}\Delta$ for $t \ge \tau$ and is zero otherwise.

In general, for any time t, let $t = n\tau + s$, where $n \ge 0$ and $0 \le s < \tau$. Then,

$$\begin{aligned} p(t) &= p(n\tau + s) \qquad (9) \\ &= p(n\tau)e^{Qs} \\ &= p((n-1)\tau)e^{Q\tau}\Delta e^{Qs} \\ &= p(0)(e^{Q\tau}\Delta)^n e^{Qs}. \end{aligned}$$

The term $e^{Q\tau}\Delta$ represents the one-step transition probability matrix of the DTMC, Y_n, embedded at the Markov regeneration epochs. If we assume that $c_1 = 1$ and $c_1 = 2$, then this DTMC is periodic with a period 2. Let $\Pi_0 = \lim_{n\to\infty}(e^{Q\tau}\Delta)^{2n}$ and let $\Pi_1 = \lim_{n\to\infty}(e^{Q\tau}\Delta)^{(2n+1)}$. Then for large t, if $t = 2n\tau + s$,

$$p(t) = \lim_{n\to\infty} p(2n\tau + s) = p(0)\Pi_0 e^{Qs}, \qquad (10)$$

and for large t, if $t = (2n+1)\tau + s$,

$$p(t) = \lim_{n\to\infty} p((2n+1)\tau + s) = p(0)\Pi_1 e^{Qs}. \qquad (11)$$

Figure 2 shows the processor utilization by periodic tasks, when priorities are assumed to be $p_1 > p_2 > p_3$. The values used are $\lambda_3 = 0.5, \mu_1 = 6.0, \mu_2 = 3.0, \mu_3 = 2, m_1 = m_2 = m_3 = 2$, and $\tau = 1$ and we assume that initially the processor is idle (time units are in seconds). As can be seen the probability exhibits periodic behavior as $t \to \infty$, with period $2\tau = 2$.

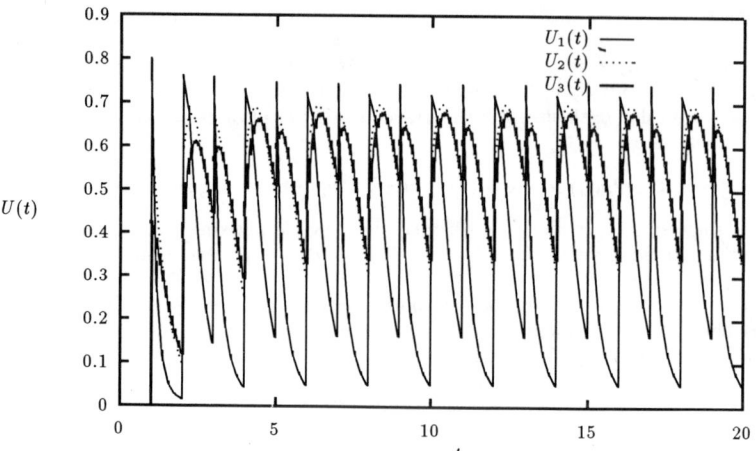

Fig. 2. Transient probability that the processor is running periodic tasks.

These numerical results confirm (and quantify) the intuitive deduction that one may make about a real-time system: that because of the periodic nature of its software tasks, the behavior of the system itself in the long run must be periodic, and must never reach a limit.

Since in the long run, utilization is periodic, we must compute time-averages if we need the long-range fraction of processing time utilized by periodic tasks. Such a quantity has been used in estimating performance of aperiodic tasks [9]. The time averaged probability vector at time t is defined as:

$$l(t) = \frac{1}{t} \int_0^t p(x)dx. \tag{12}$$

Let $t = n\tau + s, 0 \leq s < \tau$. Then we can derive $l(t)$ as follows [10]. First,

$$\int_0^t p(x)dx = \int_0^{\tau^-} p(x)dx + \int_\tau^{(2\tau)^-} p(x)dx + \int_{(n-1)\tau}^{(n\tau)^-} p(x)dx + \int_{n\tau}^{n\tau+s} p(x)dx$$

$$= \sum_{i=0}^{n-1} p(0)(e^{Q\tau}\Delta)^i \int_0^{\tau^-} e^{Qx}dx + p(0)(e^{Q\tau}\Delta)^n \int_0^s e^{Qx}dx.$$

We would like to find $\lim_{t\to\infty} l(t)$. For this, we first note that the limit of the time averaged probability of the embedded DTMC is given by [6]:

$$\lim_{n\to\infty} \frac{\sum_{i=0}^{n-1} p(0)(e^{Q\tau}\Delta)^i}{n} = \Pi,$$

where Π is given by the solution to the system of equations:

$$\Pi = \Pi e^{Q\tau}\Delta, \qquad \sum_j \Pi_j = 1.$$

For a periodic DTMC such as this one which has a period of 2, Π_j can also be computed as [6]:

$$\Pi_j = \frac{[\Pi_0]_{ij} + [\Pi_1]_{ij}}{2}.$$

Now,

$$\lim_{t \to \infty} l(t) = \lim_{n \to \infty} \left[\frac{\sum_{i=0}^{n-1} p(0)(e^{Q\tau}\Delta)^i \int_0^{\tau^-} e^{Qx} dx + p(0)(e^{Q\tau}\Delta)^n \int_0^s e^{Qx} dx}{n\tau + s} \right]$$

$$= \lim_{n \to \infty} \left[\frac{\frac{\sum_{i=0}^{n-1} p(0)(e^{Q\tau}\Delta)^i}{n} \int_0^{\tau^-} e^{Qx} dx}{\tau + s/n} \right] + \lim_{n \to \infty} \left[\frac{p(0)(e^{Q\tau}\Delta)^n \int_0^s e^{Qx} dx}{n\tau + s} \right]$$

$$= \frac{\Pi}{\tau} \int_0^{\tau^-} e^{Qx} dx. \tag{13}$$

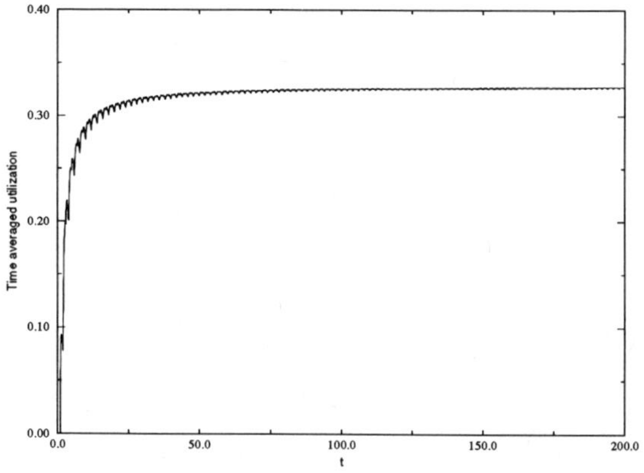

Fig. 3. Time averaged utilization by periodic tasks.

The time-averaged utilization by periodic tasks is shown in Figure 3. The long-range utilization converges to 0.33.

6 Insights and Inferences

Once the state probabilities of a model have been computed, many different performance measures can be computed by assigning appropriate weights to states and computing the weighted sum of the probabilities.

What-if analysis

Figure 4 shows the probability of being in marking M_1, which is the processor idle probability as a function of time t for the DSPN of Figure 1, assuming an initial idle processor. We show three different plots for different parameter values. $I_1(t)$ is the idle probability at time t when parameters have the following values: $\lambda_3 = 0.5, \mu_1 = 6.0, \mu_2 = 3.0, \mu_3 = 2, m_1 = m_2 = m_3 = 2$, and $\tau = 1$. $I_2(t)$

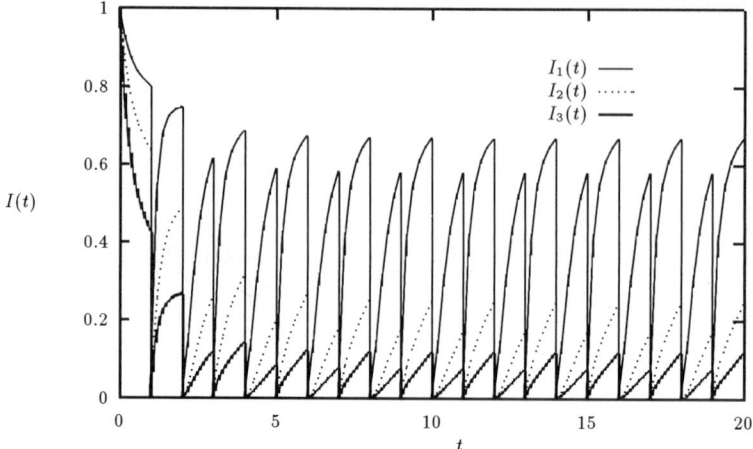

Fig. 4. Transient probability that the processor is idle.

shows the idle probability for a higher aperiodic task arrival rate and faster execution rates for the first two tasks: $\lambda_3 = 1, \mu_1 = 3.0, \mu_2 = 2.0, \mu_3 = 2, m_1 = m_2 = m_3 = 2$, and $\tau = 1$. For these values, if the aperiodic task arrival rate is increased to 1.8, the idle probability is as shown by $I_3(t)$, and is very low. Depending on the requirements of the real-time system, such low idle probability may be detrimental to the performance of the real-time system.

For the three sets of values as described above, Figure 5 shows the probability that the aperiodic task buffer is full as a function of t. Since the aperiodic task arrivals are Poisson, this is also the loss probability $l(t)$ at time t of an aperiodic task.

Figure 6 shows the loss probability for second set of values, with the buffer size for the aperiodic tasks varied. This can aid the designer in choosing the optimal buffer size, given the specification regarding the loss probability.

Advantage of using the DSPN model

It can be argued, that phase expansions can always be applied to reduce a problem to a standard Markovian modeling problem. In this case, however, the results are very sensitive to such approximations. Figure 7 shows the loss probability computed by a stochastic reward net model, where the firing time distribution of transition *timer* is approximated by an Erlang distribution with K stages. Observe that with 100 stages, the plot still does not approximate the DSPN model well. With 100 stages, the CTMC underlying the SRN had 9800 states (whereas the DSPN model has 98 states). Thus this approximation resulted in state-space explosion, while still not giving a good approximation.[3]

---

[3] This argument also implies that the assumption of phase-type execution time may also have a significant impact on the results. However, results from the literature [5] suggest that fitting phase-type distributions to service times do provide fairly good approximations. The same does not apply to approximating a periodic arrival by a phase-type renewal process, because periodic arrivals result in periodicity in the steady-state probabilities. It is this periodicity that is extremely hard to duplicate using phase-type approximations.

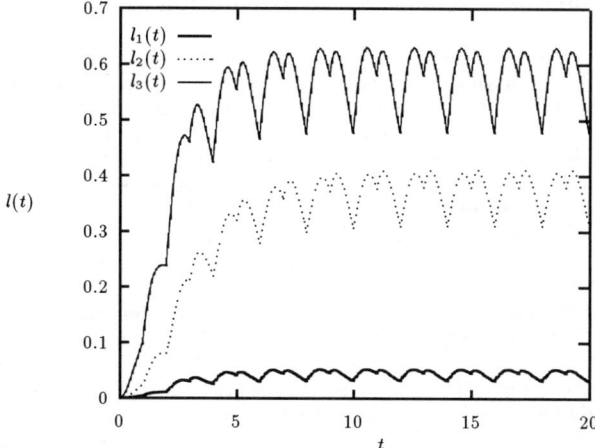

Fig. 5. Transient probability that the aperiodic task buffer is full.

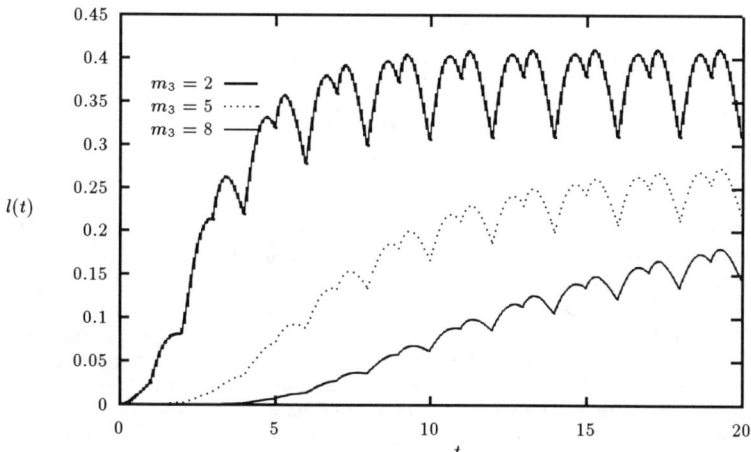

Fig. 6. Transient probability that the aperiodic task buffer is full.

7 The DSPN model for response time distribution

An important measure for tasks which have deadlines is the probability that a task meets it deadline. In this section we modify and augment our earlier model so as to compute an approximation to the response time distribution of the aperiodic task in our 3-task system. The results from the earlier section suggest that the response time distribution of a task that arrives at some time $t = n(2\tau) + s$, $0 \le s < 2\tau$ depends on s as $t \to \infty$. Thus there is no single "limiting response time distribution". In the long range, the response time distribution depends on the time of arrival to the system as measured from the beginning of the current period.

In this section we use the *tagged task approach* in computing the response time distribution given its time of arrival with respect to the time of the beginning of the current period, *and* given the fact that it is not rejected by the system.

Suppose a task arrives at time $t = 2n\tau + s$, $0 \le s < \tau$, where t is large. Since aperiodic task arrivals are Poisson, the state probability vector of the system as seen by the arriving task is $p(0)\Pi_0 e^{Qs}$ [14]. Similarly, if a task arrives at time $t = (2n+1)\tau + s$, $0 \le s < \tau$, the state probability vector of the system as seen by the arriving task is $p(0)\Pi_1 e^{Qs}$. Since we restrict our tagged jobs to those that are not rejected, these probabilities must be normalized to reflect the condition that the task is accepted by the system. Let $S(t)$ denote the state of the system as seen by a task arriving at time t that gets accepted. Then for large t,

$$\Pr[S(t) = M_i \mid t = 2n\tau + s, \#(P_3, M_i) < m_3]$$
$$= \frac{[p(0)\Pi_0 e^{Qs}]_i}{\sum_{\{j \mid \#(P_3, M_j) < m_3\}} [p(0)\Pi_0 e^{Qs}]_j}, \quad \text{if } \#(P_3, M_i) < m_3$$
$$= 0 \qquad \text{otherwise.} \qquad (14)$$

$$\Pr[S(t) = M_i \mid t = (2n+1)\tau + s, \#(P_3, M_i) < m_3]$$
$$= \frac{[p(0)\Pi_1 e^{Qs}]_i}{\sum_{\{j \mid \#(P_3, M_j) < m_3\}} [p(0)\Pi_1 e^{Qs}]_j}, \quad \text{if } \#(P_3, M_i) < m_3$$
$$= 0 \qquad \text{otherwise.} \qquad (15)$$

Let this probability be denoted by a_i.

Figure 8 shows the DSPN model for computing response time distribution of an aperiodic task. In the DSPN model, we now add a place that denotes a *tagged task*, P_{tagged}. The arc multiplicities represent the execution of this task separately, by putting 4 tokens in place P_{exec} when transition t_{tagged} fires. The arc from T_{exec} to P_{done} has variable multiplicity defined as follows: it is 1 if $\#(P_{exec}) = 4$, and is 0 if $\#(P_{exec}) \neq 4$. The inhibitor from P_3 to t_{tagged} makes sure that the aperiodic tasks that are already there when the tagged task arrives, are completed before the tagged task. When a token appears in place P_{done} the DSPN "halts", and reaches an absorbing state. This state denotes the completion of the tagged task. The initial distribution of markings of this DSPN is as follows: A marking $(M_i, \#(P_{tagged}), \#(P_{done}))$ where $M_i = (\#(P_{avail}), \#(P_1), \#(P_2), \#(P_3), \#(P_{exec}))$ has probability a_i as defined by Equations (14) and (15) above, if $\#(P_{tagged}) = 1, \#(P_{done}) = 0$ and 0 otherwise.

With this initial marking probability distribution, transient analysis can be carried out on this DSPN in a way similar to the one described in Section 5. The distribution of conditional response time, R, of the aperiodic task is given by the following:

$$\Pr[R \le x \mid \text{Time of arrival w.r.t. beginning of period and task is accepted}] =$$

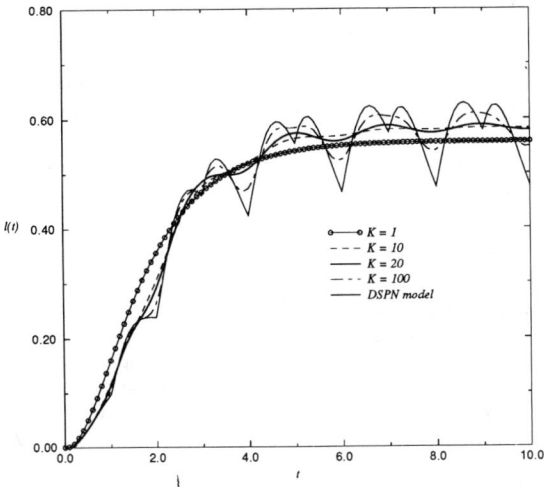

Fig. 7. Comparison of SRN model *vs.* DSPN model.

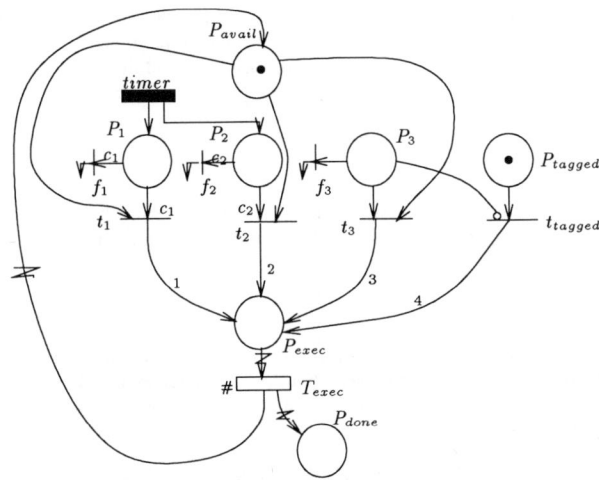

Fig. 8. DSPN model for response time distribution

$$\sum_{\{i|\#(P_{done},M_i)=1\}} p_i^{(R)}(x), \qquad (16)$$

where the superscript (R) is used to denote the fact that this is the transient probability corresponding to the response time DSPN model of Figure 8.

Figures 9 and 10 show the conditional response time distribution of an aperiodic task, for different values of s, for the first set of parameter values. Observe that the probability of missing a deadline is highest if the task arrives just after $2n\tau$. This probability is least if the task arrives at $2n\tau + s$ where s is very close to 2τ.

This evaluation can help us answering several what-if questions regarding the performance of the system. For instance, assume that the performance requirement of the aperiodic task states that in steady state at least 90% of the tasks arriving at a given offset, say $s = 0.2$, must finish within 3 seconds of arrival of the task. Figure 11 shows us the response time distribution of an arriving aperiodic task conditioned only on the fact that it arrives at a time $t = 2n\tau + 0.2$. Thus we consider the probability that it may be rejected; the distributions are therefore defective. In this case the plot tells us that the given requirement cannot be met by tasks arriving at this time if the arrival rate of the aperiodic tasks is 1.0, however a rate of 0.5 or less can be sustained.

8 Conclusions

In this paper we used Markov regenerative theory to analyze the transient behavior of a real-time system. DSPNs were used as a specification method. Transient analysis of the real-time system gave us insight into the time-dependent behavior of the real-time system. We numerically quantified the periodic behavior of the real-time system. The response-time distribution of an aperiodic task was also computed.

This technique can be implemented in a tool for real-time system analysis, to provide fast answers to performance-oriented questions. Though several simplifying assumptions were made, the paper is still a significant step forward in making models of real-time systems more accurate. Open problems that remain to be addressed include modeling of more complex scheduling policies and of non-phase-type execution times.

Acknowledgements

We wish to acknowledge the help and encouragement provided by Pam Binns of Honeywell Technology Center. We would also like to thank Dimitris Logothetis and Sandy Wang for useful discussions.

References

1. P. Binns. Design and analysis of event scheduling code in MetaH. Honeywell Internal Document, 1994.
2. E. Çinlar. *Introduction to Stochastic Processes*. Prentice-Hall, Englewood Cliffs, NJ, U.S.A., 1975.
3. H. Choi, V. Kulkarni, and K. Trivedi. Transient analysis of deterministic and stochastic Petri nets. In *Proc. of The 14th Intl. Conf. on Application and Theory of Petri Nets*, pages 166–185, Chicago, U.S.A., Jun. 21-25 1993.
4. G. Ciardo, J. Muppala, and K. Trivedi. SPNP: Stochastic Petri net package. In *Proc. Int. Conf. on Petri Nets and Performance Models*, pages 142–150, Kyoto, Japan, Dec. 1989.
5. M. A. Johnson and M. R. Taaffe. An investigation of phase-distribution moment-matching algorithms for use in queueing models. *Queueing Systems*, 8(2):129–147, 1991.
6. V. Kulkarni. *Lecture Notes on Stochastic Models in Operations Research*. University of North Carolina, Chapel Hill, U.S.A., 1990.

7. J. Kurose, D. Towsley, and C. Krishna. Design and analysis of processor scheduling policies for real-time system. In *Foundations of real-time computing: scheduling and resource management*, pages 63–90, Norwell, MA, U.S.A., 1991. Kluwer Academic Publishers.
8. P. Laplante. *Real-time systems design and analysis*. IEEE Computer Society Press, Los Alamitos, CA, U.S.A., 1993.
9. J. P. Lehocky and S. Ramos-Thuel. An optimal algorithm for scheduling soft-aperiodic tasks in fixed-priority preemptive systems. In *Proceedings of the 13th Real-Time Systems Symposium*, pages 110–123, Phoenix, Arizona, U.S.A., 1992.
10. D. Logothetis and K. Trivedi. Transient analysis of the leaky bucket rate control scheme under Poisson and ON/OFF source. In *Proc. of the 13th IEEE INFOCOMM*, Toronto, Canada, June 1994.
11. J. L. Peterson. *Petri Net Theory and the Modeling of Systems*. Prentice-Hall, Englewood Cliffs, NJ, U.S.A., 1981.
12. L. Tomek, V. Mainkar, R. Geist, and K. Trivedi. Reliability analysis of life-critical real-time systems. *Proceedings of the IEEE*, 82(1):108–121, January 1994.
13. K. S. Trivedi. *Probability & Statistics with Reliability, Queueing, and Computer Science Applications*. Prentice-Hall, Englewood Cliffs, NJ, U.S.A., 1982.
14. R. W. Wolff. Poisson arrivals see time averages. *Oper. Res.*, 30(2), Mar.-Apr. 1982.

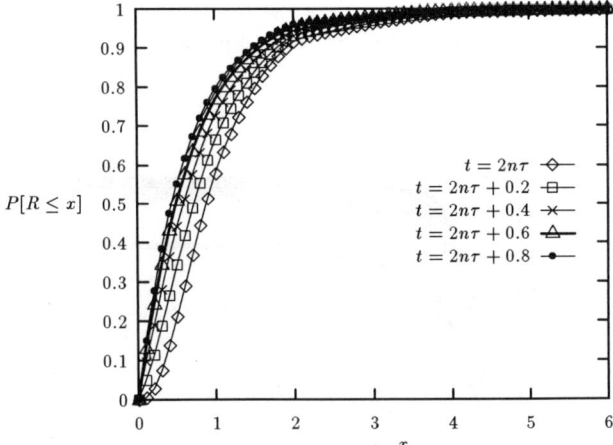

Fig. 9. Response Time Distribution For Aperiodic Task.

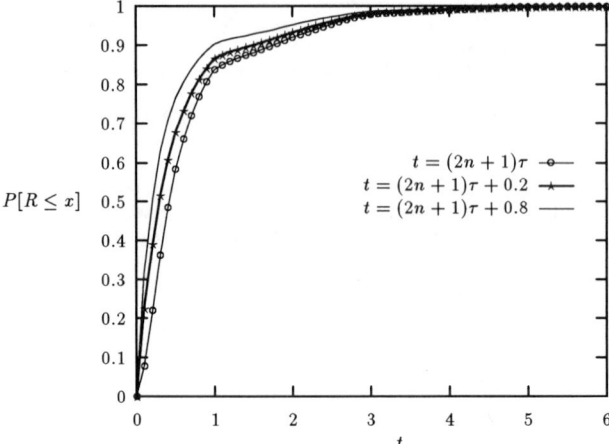

Fig. 10. Response Time Distribution For Aperiodic Task.

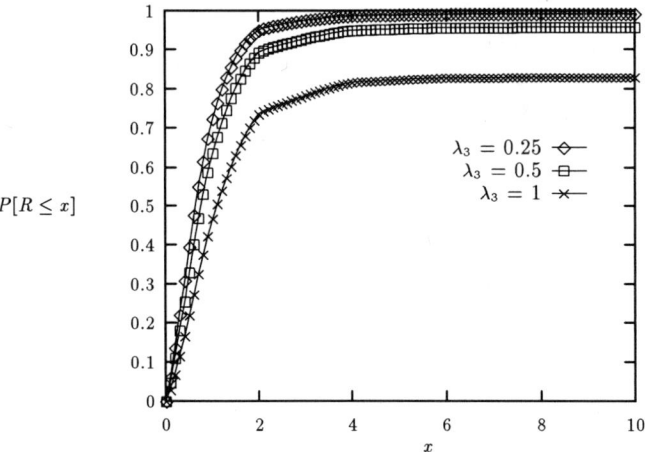

Fig. 11. Response time distribution for different aperiodic workloads.

Performance Modeling with Structured Actions

Ina Schieferdecker

GMD FOKUS(Research Institute for Open Communication Systems)
Hardenbergplatz 2, D-10623 Berlin
tel: (030) 254 99 202, fax: (030) 254 99 202, e-mail: ina@fokus.gmd.de

Abstract. The paper presents the concept of structured actions for modeling performance aspects within process-algebraic calculi. A structured action is parameterized with an interaction time, a priority, a set of requested resources, and a monitoring signal. The paper introduces the new performance-oriented LOTOS extension LOTOTIS and shows its use by specifying the Tick-Tock Case Study. Finally, it investigates the LOTOTIS upward compatibility with LOTOS.

1 Motivation

The first concern with a distributed system is its correct behavior according to the intended system functionality. The second concern is to have a proper performance with reasonable costs. While functional correctness has been considered primarily in the past, recent application of computer systems, for instance multimedia applications in broadband communication networks, showed that performance characteristics are of the same importance as the functional correctness of a distributed system. There is a need to predict performance characteristics within the design phase, to optimize system performance within the prototyping and implementation phase, as well as to control and to maintain performance during runtime.

How to develop functionally correct distributed systems starting with a formal requirement specification down to the system implementation is a well studied area. In contrast to that, it is not unusual that a system is fully implemented (or at least implemented as a prototype) and tested for functional correctness before any attempt is made to investigate its timing behavior and its performance. In case of inconvenient performance the whole development process is restarted resulting in long periods of system development, late system delivery, and high costs. In addition, there will be no guarantee that the newly designed system will have better performance than the first one.

The main problem why performance investigation are made very late in the development process, is the separation between formal specifications and performance analysis. Traditional specification techniques are lacking quantified time and other features for performance modeling. They describe only the causal ordering of events, i.e. their logical relations. On the other hand, classical performance models like queueing systems lack the description of the functional behavior, so that logical relations are often omitted. The current state of the art is to develop a performance model that is independent from the functional model. However, neither the functional behavior can be described independently of timing constraints nor the performance characteristics of a system can be derived from a performance model containing no logical relations. Furthermore, to proof that functional and performance model coincide with each other is hard if not even impossible. Henceforth, there is no possibility to investigate the functional correctness in combination with a sufficient performance. Both concerns are derived out of two independent models. The solution is to incorporate new concepts into formal specification techniques, so that the performance of distributed systems can be derived from their formal specifications. Costly re-prototyping or even re-implementation can be avoided, if performance can be predicted already within the design phase.

This paper identifies four concepts needed for performance analysis from formal specification namely quantified time, quantified nondeterminism, quantified parallelism, and monitoring. An overview and comparison of existing performance-oriented LOTOS extensions is given. Subsequently, the concept of structured actions and the new LOTOTIS approach is presented. Due to lack of space we decided to renounce the formal semantics definition. Instead, we show the applicability of our approach by specifying the Tick-Tock Case Study. Finally we investigate the upward compatibility of LOTOTIS with LOTOS.

2 Concepts Needed for Performance Modeling

The classical formal description techniques (FDTs) — LOTOS, ESTELLE, and SDL — are definitely not expressive enough to do performance analysis from formal descriptions, because they were exclusively developed for the specification of the functional behavior. In particular, all of them are lacking quantified time. Looking at the needs to express performance issues in formal specifications we identified four main concepts which are

Quantified Time Doubtless, the precondition for any performance evaluation is the possibility to express quantified time. In addition to the modeling of the causality order between events, time distances between events have to be quantified.

Quantified Nondeterminism Due to their complex behavior distributed systems are inherently nondeterministic. Their specifications represent nondeterminism by means of choices between several possible behaviors of that systems. For the sake of performance evaluation, we have to quantify these choices explicitly.

Quantified Parallelism A classic assumption of formal specification techniques is that all active behavior components — processes in case of process algebras, automata in case of finite state machines or transitions in case of Petri nets — can be executed in parallel. This corresponds to an unrestricted parallelism of the system. However, this is not adequate for performance investigations because the constraints on parallel executions have great influence on the system performance. Therefore, we need a quantified grade of allowed parallelism.

Monitoring Besides the need to describe performance-oriented aspects, there is a need to observe events during system execution — action occurrences in case of process algebras, transitions in case of finite state machines or Petri nets. This is due to the fact that events determine performance characteristics of the system under study. Hence it is necessary to make observable any event of interest — independent of whether this event is externally visible or internally hidden.

Furthermore, it is possible to define any kind of statistical inference, like computation of mean values, confidence intervals, quantiles etc. by the use of monitoring. Henceforth, event monitoring allows the unambiguous definition of performance characteristics of a system under study.

Very first approaches to close the gap between formal specification techniques and performance analysis were based on Petri net approaches. A survey is given in [2]. However, Petri net approaches are not very well accepted in practise because of lengthy and difficult to read specifications. Performance-oriented extensions of finite-state machine approaches are still marginal. On the other hand, within the field of process algebras there were developed only recently stochastic process algebras ([3], [4]). Stochastic process algebras are the basis for appropriate LOTOS extensions. They allow for compact precise behavior specifications on different levels of abstraction and in different specification styles. They provide a rich set of operators for behavior compositions and means to describe data-dependent behavior. Process algebras allow the functional as well as the structural abstraction. Hiding and renaming of actions can be used to abstract from internal details of a system (the so called black-box view). The definition of hierarchically ordered processes allow the structural abstraction by means of identifying essential subsystems and their relationships. Process algebras have a fully defined formal semantics. Last but not least, process algebras offer several equivalence notions, such as simulation, bisimulation, or testing equivalence. Under certain conditions, it is possible to prove that two behavior expressions are equivalent in a given sense. We decided to use a process-algebraic framework for the performance-oriented behavior specification of communication protocols taking into account the several advantages of process-algebraic calculi.

3 Performance-Oriented LOTOS Extensions

This section investigates existing performance-oriented LOTOS extensions. LOTOS has been standardized in 1988 [5]. It was primarily developed for the specification of communication protocols, but can also be used for the functional specification of other distributed systems. The system behavior is described in terms of actions being the atomic units of behavior, and process instances which are composed together with a set of operators. The LOTOS behavior part is based on process algebras, in particular on CCS and CSP. It offers a data type part which is based on ACT ONE. LOTOS is more and more accepted and used within industrial projects for the specification of distributed systems. In particular the need to extend LOTOS became obvious and urgent within these projects.

Firstly, let us discuss different possibilities for the introduction of time, stochastic behavior, and resources into LOTOS. [10] gives an excellent survey on the design decisions for a time concept. It differentiates between discrete vs. dense time (a minimal time unit exists or is absent), global clock vs. local clocks (either one imaginary, single clock or a couple of local clocks count the global time), time stamps vs. durations (the passage of time can either be incorporated as time consumptions in between the occurrences of actions or as time consumptions of the actions themselves), relative vs. absolute time stamps (time stamps count either relative to the enabling of an action or are absolute with respect to the global clock), maximal progress vs. action urgency (either only internal actions occur as soon as possible before any passage of time while external actions can be delayed or all actions are assumed to be urgent, i.e. as soon as an action is enabled it has to occur before any passage of time).

In order to quantify nondeterminism let us identify the potential causes of nondeterminism. At choice points several alternative behaviors are simultaneously enabled and the system has to choose one of them for future developments. While the enabling of alternative behaviors is determined by the environment, for instance by processes composed in parallel, the final decision which of the enabled behaviors is chosen is made completely internally. In classical process algebras all alternative behaviors are equally probable to be chosen, there exist no means to weight the alternatives. Another cause of nondeterminism exists within disabling expressions. In classical process algebras two behavior expressions can be combined together so that one of them may disable the other if its starting action is enabled. However, there are no means to restrict the time points, when the disabling may occur. In addition, the occurrence of the disabling action cannot be enforced in most classical process algebras, so that the disabling is unnecessarily delayed. Possible solutions are the following. Nondeterministic choices can be solved by means of action priorities and weights. Nondeterministic disablings require the specification of time points, when the disabling has to occur.

The third concept for performance analysis is quantified parallelism. The first possibility to quantify parallelism is the use of action priorities to decide on which action is executed first (assuming one imaginary processor, which has to be allocated by each action in order to be executed). However, the grade of parallelism is restricted by the number of processors and/or the number of other resources offered by the execution environment which are in general greater than one. Hence the first rule is too rigorous, so that we appreciate the use of resources as a separate modeling concept to describe the execution environment. The access to resources is again ordered by action priorities and weights.

While numerous timed LOTOS extensions have been proposed ([16], [6], [1], [11]), more elaborated LOTOS extensions towards performance evaluation are very rare (Table 1). The first performance-oriented LOTOS presented in [12] does not contain any means to describe stochastic behavior. CPT-LOTOS presented in [11] has been developed for the specification of timed systems and does not provide a concept of probabilities. Instead, it uses action priorities to solve nondeterministic choices.

To the author's best knowledge, only two LOTOS extensions are comparable to the one presented in this paper, namely [8] and [9]. The first, however, is not a proper extension of LOTOS since it excludes the disabling and enabling operator of LOTOS. In addition to a time and a priority/weight concept, the second contains the concept of random experiments to express stochastic behavior. In that respect, it is even more expressive than LOTOTIS since LOTOTIS does not contain any means to specify random variables having certain distribution functions. Both approaches [8], [9] cover quantified time and quantified nondeterminism, but the possibility to express quantified parallelism and action monitoring is not their target.

4 The Notion of Structured Actions

We have to be aware of the goals of performance evaluation in order to identify the concepts needed for performance modeling. In particular, we have to identify the aspects of distributed systems which influence the performance of these systems.

The starting point is to consider a distributed system as being composed of a number of tasks. The tasks have to be executed in parallel. The performance-influencing aspects of tasks are the following.

- Tasks are non-instantaneous, they consume time.
- Tasks allocate resources which they need for their execution.
- Task have priorities in access to these resources.

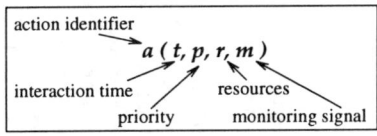

Fig. 1. The Notion of Structured Actions

In order to reflect these concepts needed for performance evaluation, we developed the concept of structured actions. We decided to structure actions with parameters for time, priorities, resources, and monitoring signals. The idea arose from the intention to make the performance-oriented modeling of distributed systems conceptually as easy as possible. Structured actions model functional

Authors	Calculus	Quantified time		Quantified nondeterminism		Quantified parallelism (Resources)	Performance Analysis
		Time	Combination of Behavior and Time	Probability	Combination of Behavior and Probability		
Rico, Bochmann [12]	STOCHASTIC LOTOS	Relative, Discrete Time	Non-instantaneous Actions, Deterministic Time Values	Probabilistic Choice	Frequencies (Weights) for Alternative Behavior	—	Performance Analysis of the Underlying Semi-Markov Model
Quemada, Frutos, Miguel [11]	CPT-LOTOS	Relative, Discrete Time	Timed Action Prefix, Timed Choice Operator	Deterministic Action Priorities	Nondeterminism is solved according to Action Priorities	—	(State Space Exploration)
Marsan, Bianco, Ciminiera, Sisto, Valenzano [8]	EXTENDED TIMED LOTOS	Relative, Discrete or Dense Time, Random Variables to Describe Passage of Time in a Probabilistic Way	Three Timer Operations: Timer, Pre-Synchronization Timer, Memory Timer	Stochastic Passage of Time, Priorities and Weights for Actions to Choose between Alternative Behavior	Time-Independent Probabilistic Choice via Pre-Synchronization Timer	—	Numerical Analysis and Simulation of the Underlying Stochastic Process
Miguel, Fernández, López, Vidaller [9]	LOTOS-TP	Relative, Discrete or Dense Time	Timed Action Prefix for Instantaneous Actions, Timed Termination	Random Experiments	Distribution Functions to Choose between Alternative Behavior	—	Next-Event Simulation of the Underlying Stochastic Process

Table 1. Performance-Oriented LOTOS Extensions

aspects of system tasks and, additionally, also their performance characteristics. The advantage of structured actions is that basically only the semantics of actions changes. There are of course several influences for the whole behavior part of the specification. The parameters of a structured action are

1. The interaction time defines the length of the synchronization period with synchronization partners, e.g. the duration of the action occurrence[1]. The interaction time makes structured actions non-instantaneous, so that they describe true-concurrent behavior.
2. The priority orders simultaneously enabled actions.
3. The set of resources has to be allocated for the execution of the interaction period[2]. The successful allocation of resources is an additional precondition for the interaction period to occur. The access to resources is adjusted by resource disciplines and action priorities.
4. The monitoring signal makes the action occurrences observable from outside. Whenever an action attached with a monitoring signal occurs, the monitoring signal offers the signal identifier and the absolute time of the action occurrence to the specification environment.

Structured actions were mainly developed during the formal semantics definition of the Timed Interacting Systems approach (TIS; ([17], [15]). Soon it became obvious, that structured actions can be applied to almost all process–algebraic calculi. The general methodology of incorporating structured actions into process algebras is given in [14].

4.1 Synchronizing Structured Actions

In order to understand the way structured actions interact (synchronize) with each other let us consider the parallel composition of two actions a which have to be executed synchronously. The structured action a is declared after the keyword **external** and has an interaction time t, priority 1, and requests two resources of type R ([R(2)]). A main characteristic of structured actions composed in parallel, independent of whether they synchronize or not, is that due to their non-instantaneous nature they may overlap in time. Structured actions model true concurrent behavior.

Let us now have a look at Fig. 2. Upon enabling, each action a is individually ready for requesting resources. However, it has to wait for its synchronization partner before really requesting the resources. Each synchronization partner becomes in general enabled at a different point in time. The first action a becomes enabled earlier than the other one. Hence there is a — not explicitly specified

---

[1] A structured action without any synchronization can be considered as a special case of synchronizing structured actions.

[2] Although resources and the resource management during system execution can be represented by additional processes and data types — this is in fact used for the formal semantics definition of LOTOTIS— we adopt the use of explicit resources as a very concise and succinct modeling feature.

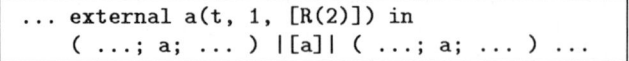

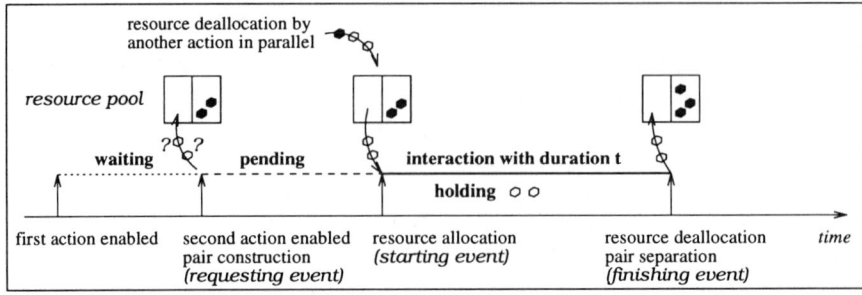

Fig. 2. Synchronization of Structured Actions

— waiting period for synchronization partners. Eventually, both actions will be enabled. Immediately, needed resources are requested (requesting event) and the actions a, willing to synchronize with each other, become pending. If all needed resources are available (two pieces of resource R), they allocate the resources and start their interaction period (starting event) immediately[3].

Resources are assigned to pending actions on an all-or-nothing principle, meaning that either all needed resources are assigned, or no assignment takes place. If possible, pending actions are assigned resources without any delay. Thus, as far as the execution environment makes it possible, interactions take place in parallel. If there are several pending actions, resources are assigned on the action priority basis. Actions having equal priority obtain resources in an order specified for each resource type by its service discipline. For time being, we restrict ourselves to the case of **fifo** (first in first out), i.e. the resources will be assigned in the same time ordering in which the actions became pending.

An interaction period takes the amount of time given by the interaction time. Allocated resources remain occupied until completion of the interaction period (finishing event). Let us repeat that an interaction period may start if and only if

1. all synchronization partners are enabled and
2. all requested resources are allocated.

An additional, not explicitly specified, waiting or pending takes place if at least one of these conditions is not satisfied. The duration of these periods can possibly extend to infinity representing deadlocks.

---

[3] We denote by synchronization of actions the whole process shown in Fig. 2. The interaction (or interaction period) refers only to the common time period of consuming the interaction time.

4.2 Independent Structured Actions

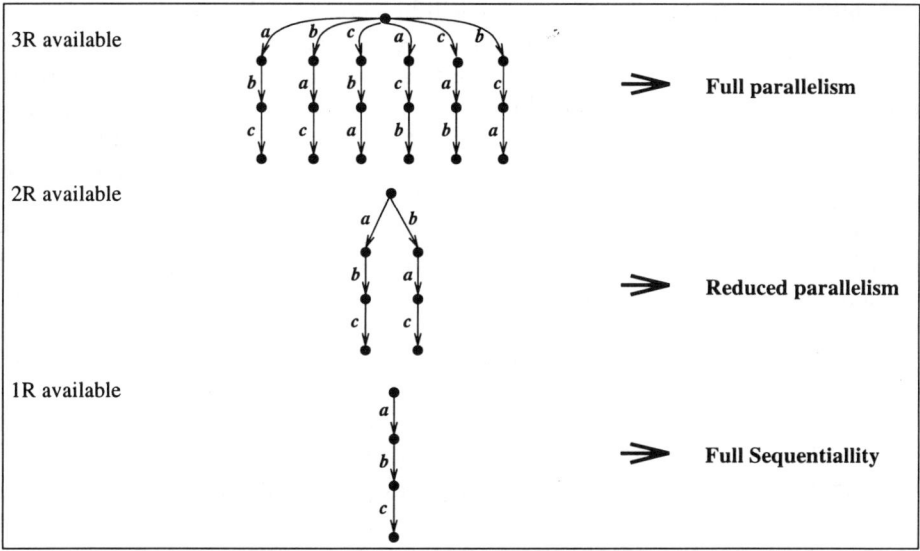

Fig. 3. Independent Structured Actions

Now, we present the modeling of full parallelism, reduced parallelism, and full sequentiality of structured actions by means of priorities and resources. Figure 3 gives three independent, instantaneous structured actions a, b, and c. Each of them needs one resource R. The figure illustrates the three interesting cases, what happens if different amounts of resource R are available. The three cases could be depicted also as 1. {a,b,c}, 2. {a,b}, {c} and 3. {a}, {b}, {c}, where actions in parenthesis denote simultaneously enabled actions, i.e. actions which can be executed in parallel. Two cases remain. The fourth case is that there is no R available: then nothing would happen except in some future time some other action in parallel to a, b, and c will deallocate resource R. The fifth case is that there are more than 3R available, however the result would be the same as for the case when 3R are available.

4.3 Performance-Oriented Operators

While structured actions can be used to model quantified time, quantified parallelism, and action monitoring, we need additional operators for the modeling of quantified nondeterminism. Although these operators are independent of the notion of structured actions, we introduce them here in order to give the reader a complete view of what we consider as quantified nondeterminism.

We need an operator to quantitatively choose between several alternative behaviors. Therefore, we decided to use a probabilistic choice operator which weights the alternative behaviors according to some probability. The expression B_1 [p] B_2 with $p \in$ [0,1] denotes that the left behavior will be chosen with probability p, while the right behavior will be chosen with probability 1-p.

Nondeterminism resulting from disabling is quantified by the use of timeout operators. A timeout operator has a time parameter t that determines the time point at which the disabling behavior expression is enabled. We distinguish between two forms of the timeout operator — the hard timeout and the soft timeout. While in the first case the disabling always occurs after t time, the disabling in the second case only occurs if the behavior expression to be disabled has not yet successfully started the interaction period of its first action. If the left behavior is fast enough to start an interaction before t time has passed it cannot be disabled anymore. In both cases, the timeout cannot occur if the behavior expression to be disabled has terminated before time t.

Please note that potential nondeterminism still exists. In particular, at time t when both the disabling behavior expression and the behavior expression to be disabled are able to execute actions with the same priority, it is undetermined which of them is chosen. Secondly, although the access to resources is adjusted by the assignment of priorities to actions and resource disciplines to resources, there is still the possibility that two equally prioritized actions request resources at the same point in time.

5 The specification language LOTOTIS

The performance-oriented specification language LOTOTIS is an LOTOS extension with structured actions, probabilistic choice operator, and timeout operators. According to our classification of LOTOS extensions LOTOTIS offers global time progress, discrete or dense time, a global clock, non-instantaneous actions, and action urgency to express quantified time. For the quantification of nondeterminism it uses probabilities to weight internal choices. Quantified parallelism is expressed by means of action priorities and resources.

Every structured action is explicitly declared by the use of the external operator (for external gates) or by the use of the hide operator (for internal actions). It is not mandatory to define all parameters of a structured action; default assumptions are zero duration, zero priority, no requested resources, and no monitoring signal. Each of the parameters can be dynamically changed during system execution by the use of the data part and appropriate value bindings. The new values for these parameters are of global significance. They are significant within the actual process and affect structured actions with the same identifier, but within different processes on the same or higher levels of the process hierarchy.

Global resources are declared on top of the specification. Additionally, each process definition may contain a resource declaration of locally available re-

```
  specification:
      specification-symbol specification-identifier formal-parameter-list
          global-type-definitions
->        [ time-declaration ]
->        [ resource-symbol resource-declarations ]
->        [ external-action-declarations ]
          behavior definition-block
      endspecification-symbol

-> time-declaration: time-symbol time-domain in-symbol
-> time-domain: discrete | dense

-> resource-declarations:
      resource-identifier '['resource-number[','resource-discipline]']'
      [ ',' resource-declarations ]

-> action-declarations:
      action-identifier '(' interaction-time [ ',' priority
                        ,'resource-request[','monitoring-signal]]]')'
      [ ',' action-declarations ]
-> external-action-declarations:
      external-symbol action-declarations in-symbol

process-definition:
  process-symbol process-identifier formal-parameter-list define-symbol
->    [ resource-declarations ]
      definition-block
  endprocess-symbol

behavior-expression:...
-> | hide-symbol action-declarations in-symbol
-> | behavior-expression '[[' probability ']]' behavior-expression
-> | behavior-expression '[[' time ']>' behavior-expression
-> | behavior-expression '[[' time '>>' behavior-expression | ...
```

Table 2. The Syntax of LOTOTIS

sources — those resources can only be accessed from process internal actions. Global resources cannot be accessed from internal actions of a locally defined process. Therefore, resources can only be allocated by actions declared on the same level as the resources themselves.

Furthermore, LOTOTIS has a probabilistic choice operator [[p]] which allows us to weight internal choices with probabilities p and 1-p for the left-hand and right-hand side behavior expression respectively, and two timeout operators — the soft timeout operator [[t]> and the hard timeout operator [[t≫. Timeout operators model a timeout after **t** time units via enabling the right-hand side behavior expression and disabling the left-hand side behavior expression. In the case of a hard timeout operator, the left-hand side behavior is disabled exactly after **t** time. In the case of a soft timeout operator, the left-hand side behavior is only disabled after **t** time if it has not started the interaction of its first structured action yet, i.e. the first structured action has successfully allocated its resources and started the interaction period with its synchronization partners. In other words, the left-hand side behavior cannot be disabled, if it is fast enough to start an interaction, before **t** time has passed.

The LotoTis syntax is given in Table 2. Since LotoTis is an extension of Lotos we only give the additional constructs that are marked with ->.

6 Specifying with LotoTis — the Tick-Tock Case Study

The Tick-Tock case study was presented in [7] as a specifically designed example for the assessment of time-extended formal description techniques. Here, it serves as an example for the use of LotoTis.

6.1 Tick-Tock Specification

The Tick-Tock system, specified in Fig. 4, is a protocol composed of two entities and an underlying service which are called **Sender**, **Receiver**, and **Service**. Each of the three components is described as a black-box with certain observable behavior. The service access points are **Se-SAP** between **User** and **Sender** represented by **s**, SAP between **Sender** and **Service** represented by **ssap**, SAP between **Service** and **Receiver** represented by **rsap**, and finally **R-SAP** between **Receiver** and **User** represented by **r**. There may exist numbers of sending users and receiving users which are identified via unambiguous addresses.

```
specification TICK-TOCK [s,r]: noexit:=
 external s, r in
 hide ssap(0), rsap(0) in    /* user primitives are instantaneous */
 ( Sender [s,ssap] ||| Receiver [r,rsap] )
 |[ ssap, rsap ]|
 Service [ssap,rsap]
 where ...
endspec
```

Fig. 4. The Tick-Tock System

6.2 Tick-Tock Service

The specification of **Service** is restricted to the interactions with **Sender** and **Receiver** through the respective SAPs. The service primitives carry a data cell as parameter and are instantaneous events, i.e. have null duration. The service is isochronous so that service primitive are only accepted at a given rate. It exists a minimal delay between cell deliveries and a minimal and a maximal transmission delay. The **Receiver** has to accept data immediately otherwise the data is lost. **Service** assumes a reliable transmission medium without cell corruption, re-orderings, and duplications. Its specification is given in Fig. 5.

```
process Service [ssap,rsap]: noexit:=
 ( From [ssap] ||| To [rsap] )
 |[ ssap,rsap]|
 Medium [ssap,rsap]
 where
 process From [ssap]: noexit:=
   internal gap(pi),         /* isochronism, at least pi time units */
           timeout(0,1) in   /* lower priority than ssap */
    gap; ssap (0,2) ?c: Cell; From [ssap]
                              /* if ssap is ready, timeout cannot occur
                                 due to its lower priority */
    [[pi>>
     timeout; From[ssap]     /* isochronism, at most pi time units */
 endproc
 process To [rsap]: noexit:=
   internal space(alpha) in
    rsap ?c: Cell; space;     /* spacing between cell deliveries */
    To [rsap]
 endproc
 process Medium [ssap,rsap]: noexit:=
   internal delay(tau_min),   /* transmission delay */
           timeout(0,1) in    /* lower priority than rsap */
    ssap ?c: Cell;
    let t:TIME= in[tau_min,tau_max] in
    delay(t);                 /* undetermined delay */
    ( rsap (0, 2) !d; Medium [ssap,rsap]
                      /* immediate acceptance, if rsap is ready,
                         timeout cannot occur due to its lower priority */
      [[0>>
       timeout; Medium [ssap,rsap])/* data loss, if receiver not ready */
 endproc
endproc
```

Fig. 5. The Tick-Tock Service

The behavior of **Service** is modeled as follows: The immediate cell acceptance by **Receiver** is modeled by introducing an internal action **timeout**, which forces the **Medium** to timeout immediately if the **Receiver** is not ready to accept the cell. Otherwise, due to the higher priority of **rsap** in comparison to **timeout**, **Receiver** can always accept a cell whenever it is able to do so. The urgency for cell deliveries, that is effected by two internal delay, is constrained by the transmission delay **delay** and the delay between successive deliveries **space**. Both determine the instants of cell arrivals. We express isochronism by two time constraints: the first claims that there are at least **pi** units in between successive cell acceptances and the second claims that there are at most **pi** units in between. This enforces cell acceptances at precise, punctual instants. The undetermined delay of the internal action **delay** is reflected by dynamically setting its interaction time in between the minimal and maximal value of the delay.

6.3 The Tick-Tock Example extended with other Aspects

So far we only considered the time-critical behavior of the Tick-Tock system. However, as argued at the beginning of this paper not only time-dependencies, but also quantified nondeterminism and quantified parallelism influence the performance of a given system. Subsequently we give additional performance-oriented aspects for the Tick-Tock system. More realistic assumptions for the transmission medium which is used by **Service** are, that it is an unreliable medium with cell corruption, re-orderings and duplications. The unreliability of a transmission medium can be modeled via probabilistic choice operators, where the probabilities correspond to the cell loss rate or the cell corruption rate of the transmission medium respectively. We can specify it as presented in Fig. 6.

```
process Medium [ssap,rsap]: noexit:=
  ...
  ( ( rsap (0, 2) !d; Medium [ssap,rsap]
    [q]
    rsap (0, 2) !corrupted(d); Medium [ssap,rsap] )
                    /* data corruption with probability 1-q */
    [[0>>
    timeout; Medium [ssap,rsap])/* data loss, if receiver not ready */
    [[p]]
    Medium [ssap,rsap]/* data loss during transmission with prob. 1-p */
endproc
```

Fig. 6. The Tick-Tock System with Unreliable Transmission Medium

Cell re-orderings and cell duplications have to be modeled via additional internal mechanisms for the transmission medium, which describe the way cells are re-ordered and/or duplicated. This can be done in a pure functional manner and requests only for quantifications how often cell sequences are corrupted. Please note that if we assume an unreliable transmission medium the Tick-Tock system has to contain some additional error recovery mechanisms which will change the specifications for **Sender** and **Receiver** given above.

Additionally, we can investigate the influence of processor availability on the system performance. Although the Tick-Tock system assumes arbitrary numbers of sending users and receiving users, nothing was said about the processing capacity of the Tick-Tock system. Implicitly we assumed that any component of the system, i.e. users, the sender, service, and receiver, are able to execute their functionalities at need — there are no restrictions for the system parallelism at all. However, if we assume a given number of processors and define a mapping from the system components onto these processors we will get more interdependencies for them. The mapping between structured actions and the processors can be defined by identifying classes of processors and assigning to every structured action one specific class, from which it requests one processor. Interesting questions are (1) what is the maximal performance increase that can be achieved by use of additional processors or (2) what is an optimal mapping of the system components to the processors for best performance. For illustration we give in Fig. 7 one example for a Tick-Tock system assuming that the sending users with the sender, the service itself, and the receiving users with the receiver share one processor class, respectively. The number of processors per processor class can be parameterized.

```
specification TICK-TOCK [s,r](c1,c2,c3: Nat): noexit:=
 resources p1(c1),p2(c2),p3(c3) in           /* three processor classes */
 external s(0,0,[p1]),r(0,0,[p2]) in         /* sending users and Sender */
                 /* share p1, receiving users and Receiver share p2 */
 hide ssap(0,0,[p3]),rsap(0,0,[p3]) in
                                 /* ssap, rsap, and Service share p3 */
 ( Sender [s,ssap] ||| Receiver [r,rsap] )
 |[ ssap, rsap ]|
 Service [ssap,rsap]
 where ...endspec
```

Fig. 7. The Tick-Tock System with Restricted Set of Processors

The main outcome of the study is that LOTOTIS is able to model all time-dependent aspects which were identified to be crucial for the correct specification of time-dependent behavior. It allows for very compact and precise description of time dependencies. Data dependent behavior is an elegant mean to describe dynamic behavior aspects during system execution. But even more, the purely time-oriented Tick-Tock specification has been directly and very easily extended with additional performance-oriented aspects. The concept of resources allows to describe in a concise manner the influence of the execution environment for the grade of parallelism.

6.4 Upward Compatibility with LOTOS

This section shows that LOTOTIS is an upward compatible extension of LOTOS.

1. Every LOTOS specification is syntactically also a LOTOTIS specification.
2. The semantics of a LOTOS specification remains unchanged if this specification is interpreted by the LOTOTIS semantics rules. The LOTOS semantics is preserved.
3. The semantics of a LOTOTIS specification is a refinement of the semantics of its underlying LOTOS specification. It is a subset of the behavior of the underlying LOTOS specification and does not contradict.

The aspect of syntax compatibility is easily shown by checking the formal syntax of LOTOS and of LOTOTIS. We do not explore more on the semantics preservation since its proof is technical along the lines of the LOTOTIS semantics definition given in [13]. Now, let us consider the refinement relation.

A refinement relation, also called simulation relation, is a relation between transition systems. Two transition systems T_1 and T_2 stand in a refinement relation $T_1 \preceq T_2$, read as T_1 refines T_2, if all transitions that T_1 is able to execute can also be executed by T_2. A classic definition for a refinement relation is the following one. Let $T_i = < S_i, A, \longrightarrow, s_0^i >, i = 1, 2$ be two transition system over the same language and with identical set of actions. A relation $R \subseteq S_1 \times S_2$ is a refinement relation if and only if: if $(s_1, s_2) \in R$ then for all $s_1' \in S_1$ with $s_1 \xrightarrow{a} s_1'$ there is a $s_2' \in S_2$ with $s_2 \xrightarrow{a} s_2'$ and $(s_1', s_2') \in R$. T_1 refines T_2, i.e. $T_1 \preceq T_2$, if and only if there is a refinement relation $R \subseteq S_1 \times S_2$ with $(s_0^1, s_0^2) \in R$.

The difficulties in defining a refinement relation for LOTOTIS and LOTOS are due to two facts:

1. we consider transition systems of different languages, and
2. the sets of actions are different, too.

Therefore, we introduce a comparison relation on actions that allows us to interrelate LOTOTIS and LOTOS transitions. Firstly, let us give some general definitions concerning the transition relation of a transition system. Let $T = < \mathcal{B}, \mathcal{A}, \longrightarrow, B_0 >$ be a structured labeled transition system. Furthermore let $B, B' \in \mathcal{B}$, $a_1, \ldots, a_n \in \mathcal{A}$, and $a \in \mathcal{A} \setminus \mathcal{A}^*$. We define

Transition series: $B \xrightarrow{a_1, \ldots, a_n} B'$ if and only if there is $B_1, \ldots, B_{n+1} \in \mathcal{B}$ with $B = B_1, B' = B_{n+1}$, and $B_i \xrightarrow{a_i} B_{i+1}$ for all $i = 1, \ldots, n$,

Weak transition series up to $\mathcal{A}^*$: $B \xRightarrow{b}_{\mathcal{A}^*} B'$ for $\mathcal{A}^* \subseteq \mathcal{A}$ if and only if $B \xrightarrow{t\,b} B'$ with $t \in \bigcup_{k \in \mathbb{N}} \mathcal{A}^{*^k}$, where $\mathcal{A}^{*^k}$ denotes the set of finite sequences of actions in $\mathcal{A}^*$ of length k.

The definition given in Def. 1 resembles directly the ideas of the "classic" definition of a refinement relation.

Definition 1. Let $T_M = <\mathcal{B}_M, \mathcal{A}_M, \longrightarrow_M, B_0^M>$ and $T_N = <\mathcal{B}_N, \mathcal{A}_N, \longrightarrow_N, B_0^N>$ be two transition systems of different languages. Let $=_\mathcal{A} \subseteq \mathcal{A}_M \times \mathcal{A}_N$ be a comparison relation between the action of T_M and T_N. Furthermore, let $\mathcal{A}_M^* = \{a \mid a \in \mathcal{A}_M, \forall a' \in \mathcal{A}_N : \neg (a =_\mathcal{A} a')\}$ and $\mathcal{A}_N^* = \{a \mid a \in \mathcal{A}_N, \forall a' \in \mathcal{A}_M : \neg (a =_\mathcal{A} a')\}$ be the finite sets of incomparable actions of $\mathcal{A}_M$ and $\mathcal{A}_N$, respectively.

A relation $R_{=_\mathcal{A}} \subseteq \mathcal{B}_M \times \mathcal{B}_N$ is a **refinement relation with respect to** $=_\mathcal{A}$ if and only if:

if $(B_M, B_N) \in R$ then
whenever $B_M \stackrel{a}{\Longrightarrow}_{\mathcal{A}_M^*} B_M'$ with $a \notin \mathcal{A}_M^*$ then for some $B_N' : B_N \stackrel{b}{\Longrightarrow}_{\mathcal{A}_N^*} B_N'$ with $a =_\mathcal{A} b$ and $(B_M', B_N') \in R$.

T_M is a **refinement** of T_N with respect to $=_\mathcal{A}$, denoted by $T_M \preceq_{=_\mathcal{A}} T_N$, if there is a refinement relation $R \subseteq \mathcal{B}_M \times \mathcal{B}_N$ with respect to $=_\mathcal{A}$ and with $(B_0^M, B_0^N) \in R$. •

We want to show that a LOTOS action **a** is refined by a structured LOTOTIS action **a**. The semantics of LOTOTIS defines the GENIUS action **a !at_s** which corresponds to the start of the interaction period, to represent the occurrence of the structured LOTOTIS action **a**. Hence, we define the refinement relation based on the GENIUS actions **a !at_s** and the LOTOS actions **a**. Therefore, the comparison relation contains pairs of the form $(a[at\_s], b)$ with $name(a) = name(b)$. For more information refer to [13].

In order to show the refinement relation between LOTOTIS and LOTOS, we offer general transformation rules that allow one to derive performance-oriented specifications from existing LOTOS specifications (Table 3). Structuring external and/or internal actions comprises the transformation of defining action parameters for a given action — defining its interaction time, its priority, its resources, and/or its monitoring signal. It assumes that the time domain and the used resource are properly declared.

Quantifying disablings is the transformation of defining a time parameter for a disabling operator. This reduces the possibilities, when the disruption can occur.

The third transformation is the parameterization of choice operators that weights the alternatives of the choice expression with probabilities for their occurrences.

The following theorem asserts that all three transformations provides refinement rules from LOTOS to LOTOTIS.

Theorem 2. *Structuring actions, quantifying disablings, and quantifying choices in a LOTOS specification are refinement rules. They yield LOTOTIS specifications, which are refinements of the original LOTOS specification.*

	LOTOS construct	LOTOTIS construct	
Structuring Actions	External action a $\longmapsto$ implicitly declared as gate of the LOTOS specifications	time ... in resource R[...],... in external a(t) in $\longmapsto$ ... external a(t,p) in $\longmapsto$ ... external a(t,p,r) in $\longmapsto$ ... external a(t,p,r,m) in	or or or
	Internal action a $\longmapsto$ hide a in	time ... in resource R[...],... in hide a(t) in $\longmapsto$ ... hide a(t,p) in $\longmapsto$ ... hide a(t,p,r) in $\longmapsto$ ... hide a(t,p,r,m) in	or or or
Quantifying Disablings	[> $\longmapsto$ [[t]> $\longmapsto$ [[t≫		or
Quantifying Choices	[] $\longmapsto$ [[p]]		

Table 3. The Refinement Rules from LOTOS to LOTOTIS

The refinement relation between LOTOTIS and LOTOS leads us additionally to a sufficient criteria when a LOTOTIS specification has a finite state space. Finiteness of LOTOTIS specifications is a crucial precondition for a number of verification methods that are used to reason about properties of the specified system.

Theorem 3. *Let L be a guarded LOTOTIS specification. If the underlying LOTOS specification L_{LOTOS} of L has a finite state space, then L also has a finite state space.* •

Due to lack of space the paper cannot offer proofs for these theorems, please refer to [13].

7 Conclusions

Firstly, we motivated the need for new specification techniques which allow for the functional and performance-oriented behavior description of distributed systems based on one formalism. We identified four main concepts that have to be introduced into functional specification techniques, namely quantified time, quantified nondeterminism, quantified parallelism, and monitoring. Afterwards we compared recent proposals and identified their advantages and drawbacks.

The main contribution of this paper is the presentation of the concept of structured actions which provides means for the specification of performance-oriented behavior in a very compact and concise manner. Based on the concept of structured actions, we presented an upward compatible extension with LOTOS that is called LOTOTIS. Since LOTOTIS has an operational semantics we can verify functional properties like deadlock freeness or reachability of a LOTOTIS specification by the use of classical methods, such as state-space exploration techniques. Furthermore, LOTOTIS offers the notion of bisimulation which can be used to investigate the equivalence of two LOTOTIS specifications.

We demonstrated the applicability of our approach by specifying the Tick-Tock case study. We proposed additional performance-oriented aspects of this case study that should be investigated in the future. At the end, we investigated the upward compatibility of LOTOTIS with LOTOS and gave three refinement rules that can be used to derive a performance-oriented specification from existing LOTOS specifications. The upward compatibility ensures that the functional behavior of the resulting LOTOTIS specification is a refinement of the original LOTOS behavior.

References

1. T. Bolognesi, F. Lucidi, and S. Trigila. Towards full timed LOTOS. In *1st AMAST Workshop on Real-Time Systems*, 1993.
2. G. Ciardo, R. German, and C. Lindemann. A characterization of the stochastic process underlying a stochastic Petri net. In *PERFORMANCE '93, Rome, Italy*, 1993.
3. N. Götz, U. Herzog, and M. Rettelbach. Multiprocessor and distributed system design: The integration of functional specification and performance analysis using stochastic process algebras. In *PERFORMANCE '93, Rome, Italy*, 1993.
4. J. Hillston. PEPA: A performance enhanced process algebra. Technical Report CSR-24-93, University of Edinburgh, 1993.
5. ISO. LOTOS - a formal description technique based on the temporal ordering of observational behaviour. ISO/IEC 8807, 1988.
6. G. Leduc and L. Léonard. A timed LOTOS supporting a dense time domain and including new timed operators. In *FORTE V*, pages 99–114, 1992.
7. L. Léonard, G. Leduc, and A. Danthine. The tick-tock case study for the assessment of timed FDTs. Esprit Project 5341/Sector OBS, OSI 95, 1993.
8. M.A. Marsan, A. Bianco, and et. al. Integrating performance analysis in the context of LOTOS-based design. In *MASCOTS'94*, pages 292–298, 1994.
9. C. Miguel, A. Fernández, J.López, and L. Vidaller. A LOTOS based performance evaluation tool. *Computer Networks and ISDN Systems*, 25(7):791–814, 1993.
10. X. Nicollin and J. Sifakis. An overview and synthesis on timed process algebras. In *CAV'91*, pages 1–21, 1991.
11. J. Quemada, D.d. Frutos, and C. Miguel. On the design of timed systems. In *1st AMAST Workshop on Real-Time Systems*, 1993.
12. N. Rico and G.v. Bochmann. Performance description and analysis for distributed systems using a variant of LOTOS. In *PSTV XI*, 1999.
13. I. Schieferdecker. Performance-oriented specification of communication protocols and verification of deterministic bounds on their qos characteristics. Preliminary Version of the PhD Thesis.
14. I. Schieferdecker. Structuring actions for performance-enhanced specifications. Submitted to Journal on Real-Time Systems.

15. I. Schieferdecker and A. Wolisz. Operational semantics of timed interacting systems: an algebraic performance oriented formal description technique. 92/19, Technical University Berlin, 1992.
16. W.H.P. van Hulzen, P.A.J. Tilanus, and H. Zuidweg. LOTOS extended with clocks. In *FORTE II*, pages 179–191, 1990.
17. A. Wolisz. A unified approach to formal specification of communication protocols and analysis of their performance. *Journal of Math. Modelling and Simulation in Systems Analysis*, (1993)(10), 1993.

Transient Analysis of Deterministic and Stochastic Petri Nets by the Method of Supplementary Variables *

Reinhard German

Technische Universität Berlin, Prozeßdatenverarbeitung und Robotik,
Franklinstr. 28/29, 10587 Berlin, Germany

Abstract. This paper addresses the transient analysis of stochastic Petri nets with deterministic and exponentially distributed firing times. Using the method of supplementary variables general state equations are derived which are valid under certain structural conditions. A numerical technique for the solution of the general state equations is proposed. Furthermore, a special case is identified, for which an efficient and easy to implement solution is possible. This technique can be used for the approximate solution in more general cases. Numerical examples demonstrate the use of the different algorithms.

1 Introduction

Stochastic Petri nets (SPNs) [1] represent a graphical method for the modeling of discrete event systems like computer systems, communication systems, and manufacturing systems. The stochastic extensions to the pure Petri net formalism allow to model and evaluate the performance and dependability of these systems. Most commonly, the transitions may fire either without consuming time or after an exponentially distributed delay. In order to allow more general firing times, Ajmone Marsan and Chiola defined a new class of SPNs, refered to as *deterministic and stochastic Petri nets* (DSPNs) [3]. In DSPNs transitions may fire without consuming time (*immediate transitions*), after a deterministic time (*deterministic transitions*), or after an exponentially distributed time (*exponential transitions*). The mixing of deterministic and stochastically distributed delays is very appealing since the formalism is well suited for the modeling of technical systems in which activities with constant and with unknown durations may occur (e.g. real-time systems where faults may occur randomly, refered to as *responsive systems* in [20]).

In order to obtain quantitative results from the model, the stochastic process which is underlying a DSPN has to be evaluated either by analysis or by simulation. As it was shown in [3], the steady-state analysis of DSPNs is possible, if in each marking no more than one deterministic transition is enabled. The analysis method is based on the definition of an emdedded Markov chain

---

* This work was supported by the Siemens Corporate Research and Development

at certain instants of time. These instants of time represent *regeneration points* of the stochastic process (i.e. the process is memoryless at these instants). This principle was first developed by Kendall in the context of queueing systems and later extended for more general stochastic processes [8, 15, 9]. In [12] it was shown, that also the method of *supplementary variables*, initially proposed by Cox [10], can be used for the steady-state analysis of DSPNs under similar structural restrictions. In that approach the discrete state description of the stochastic process is supplemented by age variables which represent the relevant information about the history. Therfore a mixed discrete and continuous state Markov process is defined for which the Kolmogorov state equations can be derived and numerically analyzed. The approaches based on regeneration points and on supplementary variables lead to similar, but slightly different state equations. Other authors have also examined the steady-state solution of DSPN-like formalisms [6].

Recently, also the transient analysis of DSPNs was addressed in the literature. Choi, Kulkarni, and Trivedi derived state equations which describe the transient behavior of the stochastic process underlying a DSPN [7]. The approach is based on regeneration points and leads to complex equations. Although numerical Laplace transform inversion could be used for the numerical solution, this seems not to be a good choice due to numerical instabilities and costs. Therefore, Logothetis and Trivedi considered several special structures, for which a more efficient numerical solution of the state equations is possible [16, 17, 18]. However, no feasible general method for the solution of the state equations is known. Bobbio and Telek also use an approach based on regeneration points for obtaining state equations in case of more general firing policies of the transitions [5].

In this paper we use the method of supplementary variables for the derivation of state equations which describe the transient behavior of a DSPN under certain structural conditions. The state equations differ from the equations which are obtained by the method of regeneration points and can be used for the numerical computation of the transient probabilities. In the special case that in each marking a deterministic transition with the same firing time is enabled, an efficient and easy to implement solution is possible. In the more general case an iterative scheme using Runge-Kutta standard methods is proposed. Furthermore, it is proposed to use the solution technique of the special case for the approximate transient solution of more general DSPNs. The purpose of this work is to continue the work in [7, 16, 17, 18]. The special case identified in this paper is a generalization of the case considered in [16]. In that reference solution formulas were given for a model of a leaky-bucket rate control scheme, which is similar to a D/M/1/K queueing system.

The remainder of this paper is organized as follows. In Section 2 it is shown how general state equations describing the transient behavior of DSPNs can be derived by the method of supplementary variables. Subsequently, Section 3 addresses the solution of the state equations. Numerical examples are given in Section 4 and final conclusions in Section 5.

2 Derivation of the State Equations

In this section we derive general state equations describing the transient behavior of the stochastic process underlying a DSPN. First the considered class of stochastic Petri nets is introduced. Then the state equations are derived in a general matrix form. Subsequently, it is shown that these equations reduce to the equations which are known for a M/D/1/K queueing system.

2.1 The Considered Class of Stochastic Petri Nets

We consider the class of *deterministic and stochastic Petri nets* (DSPNs) [3]. A DSPN consists of *places*, *transitions*, and *arcs*. Places may contain undistinguishable *tokens*. The transitions may fire without consuming time (*immediate transitions*), after a deterministic time (*deterministic transitions*), or after an exponentially distributed time (*exponential transitions*). The arcs are divided into *input*, *output* and *inhibitor arcs*. The vector representing the number of tokens in each place is refered to as *marking*. Depending on the arcs transitions may be *enabled* in a marking. Enabled transitions may *fire* leading to a marking change. The customary enabling and firing rules are adopted as e.g. described in [2]. Additionally, it is assumed that the deterministic transitions have the firing policy *race with enabling memory* [2].

General equations describing the transient behavior of a DSPN are derived. The equations are subject to the following three restrictions:

1. in each marking no more than one deterministic transition may be enabled,
2. all deterministic transitions have the same firing time τ,
3. a deterministic transition must not be preempted.

All restrictions could be relaxed leading to more complex state equations. However, relaxion of restriction 1 leads to very complex state equations which seem to be unfeasible for analysis, relaxion of restriction 2 leads just to a more verbose notation, and relaxion of restriction 3 seems also possible.

2.2 General Form of the State Equations

The tangible markings of a DSPN constitute the states of an underlying stochastic process. We assume that the number of tangible markings is finite and, thus, the state space $\mathcal{S}$ can be enumerated:

$$\mathcal{S} = \{0, \ldots, K\}. \tag{1}$$

The state space can be partitioned into the sets $\mathcal{S}^E$ and $\mathcal{S}^D$. In $\mathcal{S}^E$ only exponential transitions are enabled and in $\mathcal{S}^D$ a deterministic transition is enabled. The aim of our analysis is to compute the transient state probabilities of the stochastic process, denoted as $p_n(t)$, for $0 \leq n \leq K$ and for $t \in \mathbb{R}^+$.

Since a deterministic firing time is not memoryless, the evolution of the underlying stochastic process depends on the history. Supplementary variables, as proposed by Cox [10], can be used to represent the relevant information about the history in the current state description. Accordingly, a continuous-state Markov process is defined for which the Kolmogorov state equations can be derived. The method of supplementary variables is valid for general firing time distributions, and in case of deterministic firing times considerable simplifications of the equations are possible. Let X be the age variable representing the time since the deterministic transition has become enabled. Then, let $p_n(t,x)$ denote the density function of the age in state n:

$$p_n(t,x) = \frac{d}{dx} Pr\{\text{state } n \text{ at time } t, X \leq x\}, \text{ for } n \in \mathcal{S}^D. \qquad (2)$$

The state equations can be derived for these quantities. Recall that τ denotes the firing time of the deterministic transitions. Having obtained the $p_n(t,x)$, the $p_n(t)$ are given by integration:

$$p_n(t) = \int_0^\tau p_n(t,x)\, dx, \text{ for } n \in \mathcal{S}^D. \qquad (3)$$

In order to give the state equations in matrix notation, the following vectors and matrices are defined:

- vectors: $\mathbf{p}^E(t)$, $\mathbf{p}^D(t)$, and $\mathbf{p}^D(t,x)$
- matrices: $\mathbf{Q}^{E,E}$, $\mathbf{Q}^{E,D}$, $\mathbf{Q}^D$, $\mathbf{\Delta}^E$, and $\mathbf{\Delta}^D$

All vectors are row vectors of dimension $K+1$ and all matrices are square matrices of dimension $K+1$. The entries of $\mathbf{p}^E(t)$ corresponding to states of $\mathcal{S}^E$ are given by the transient state probabilities in these states, all other entries are set to zero. The entries of $\mathbf{p}^D(t)$ and $\mathbf{p}^D(t,x)$ are analogously defined:

$$p_n^E(t) = \begin{cases} p_n(t) & \text{if } n \in \mathcal{S}^E \\ 0 & \text{otherwise} \end{cases}, \qquad (4)$$

$$p_n^D(t) = \begin{cases} p_n(t) & \text{if } n \in \mathcal{S}^D \\ 0 & \text{otherwise} \end{cases}, \quad p_n^D(t,x) = \begin{cases} p_n(t,x) & \text{if } n \in \mathcal{S}^D \\ 0 & \text{otherwise} \end{cases}. \qquad (5)$$

The matrices $\mathbf{Q}^{E,E}$ and $\mathbf{Q}^{E,D}$ represent the rates of all exponential state transitions outgoing from states of $\mathcal{S}^E$. The sum of both matrices is the generator matrix of a continuous-time Markov chain (CTMC). $\mathbf{Q}^{E,E}$ comprises all non-zero entries leading to states of $\mathcal{S}^E$ and $\mathbf{Q}^{E,D}$ comprises all non-zero entries leading to states of $\mathcal{S}^D$. The matrix $\mathbf{Q}^D$ represents the generator matrix of the CTMC of all possible exponential state transitions outgoing from states of $\mathcal{S}^D$. Note that no exponential transition is possible from states of $\mathcal{S}^D$ to states of $\mathcal{S}^E$ due to restriction 3 in Section 2.1. Finally, the matrices $\mathbf{\Delta}^E$ and $\mathbf{\Delta}^D$ contain the switching probabilities after the firing of a deterministic transition. The switching probabilities represent the probabilities of immediately switching to other states after a deterministic transition has fired. The switching is caused by the firing of the deterministic transitions themselves and by possible firings of the immediate transitions. Δ^E and Δ^D contain entries leading to states of $\mathcal{S}^E$ and $\mathcal{S}^D$, respectively.

The transient behavior of the system can now be described by the following equations. In case a deterministic transition is enabled, i.e. for states of $\mathcal{S}^D$, the evolution for $0 < x \leq \tau$, $t > 0$ is given by:

$$\left(\frac{\partial}{\partial t} + \frac{\partial}{\partial x}\right) \mathbf{p}^D(t, x) = \mathbf{p}^D(t, x) \cdot \mathbf{Q}^D. \tag{6}$$

Equation (6) is a system of *partial differential equations* (PDEs) of first order with constant coefficients. Equation (6) reflects the fact that states in which a deterministic transition is enabled and with an age variable greater than zero can only be reached from states in which the deterministic transition was already enabled. For states of $\mathcal{S}^E$ the evolution for $t > 0$ is described by:

$$\frac{d}{dt}\mathbf{p}^E(t) = \mathbf{p}^E(t) \cdot \mathbf{Q}^{E,E} + \mathbf{p}^D(t, \tau) \cdot \Delta^E. \tag{7}$$

Equation (7) is a system of *ordinary differential equations* (ODEs) with constant coefficients which is coupled with the system of PDEs by means of the values of $\mathbf{p}^D(t, \tau)$. The change of the state probabilities is given by the possible exponential state transitions (expressed by the term $\mathbf{p}^E(t) \cdot \mathbf{Q}^{E,E}$) and by the firing of the deterministic transitions (expressed by the term $\mathbf{p}^D(t, \tau) \cdot \Delta^E$). Note that the number of ODEs corresponds to the number of states in which only exponential transitions are enabled (i.e. number of states of $\mathcal{S}^E$).

Furthermore, we assume that the initial state is known and that no deterministic transition was previously enabled (i.e. the age variables are initially set to zero). Hence the following *initial conditions* are known:

$$\mathbf{p}^E(0) \text{ given}, \ \mathbf{p}^D(0) \text{ given}, \ \text{and } \mathbf{p}^D(0, x) = \mathbf{0}, \text{ for } x > 0. \tag{8}$$

Additionally, the following *boundary conditions* can be derived for $t > 0$:

$$\mathbf{p}^D(t, 0) = \mathbf{p}^D(t, \tau) \cdot \Delta^D + \mathbf{p}^E(t) \cdot \mathbf{Q}^{E,D}. \tag{9}$$

Equation (9) reflects the fact that states in which a deterministic transition is enabled and with an age variable equal to zero can only be reached by exponential state transitions from states of $\mathcal{S}^E$ (expressed by the term $\mathbf{p}^E(t) \cdot \mathbf{Q}^{E,D}$) or by firings of deterministic transitions themselves (expressed by the term $\mathbf{p}^D(t, \tau) \cdot \Delta^D$). The vector $\mathbf{p}^D(t)$, $t \geq 0$ is then given by the following vector *integral equation*:

$$\mathbf{p}^D(t) = \int_0^\tau \mathbf{p}^D(t, x)\, dx. \tag{10}$$

The transient behavior of the stochastic process underlying a DSPN is thus described by the system of PDEs (6) and the system of ODEs (7), subject to the initial (8) and boundary conditions (9). The transient state probabilities for states of $\mathcal{S}^E$ are already specified by these equations, the transient probabilities of states of $\mathcal{S}^D$ can be obtained by the integral equation (10).

2.3 An Example: M/D/1/K Queueing System

In this subsection we consider a queueing system with Poisson arrival, a finite waiting room of K places and a single deterministic server, customarily refered to as a M/D/1/K queueing system. First, the sets, vectors, and matrices defined in the last section are derived for that system. Then single state equations are obtained by inserting the vectors and matrices into the general state equations. The resulting equations were already described elsewhere and can be compared with [10] and [13].

Figure 1 shows the state graph of the underlying stochastic process. The number of each state corresponds to the number of customers in the system. States in which the deterministic service is active are supplemented by the age variable x. Exponential state transitions are drawn as thin lines labeled with the rate λ and deterministic state transitions are drawn as thick lines labeled with the deterministic firing time τ.

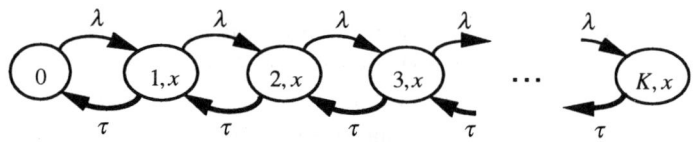

Fig. 1. Stochastic process underlying a M/D/1/K queueing system

The sets of states are given by:

$$S = \{0, \ldots, K\}, \quad S^E = \{0\}, \quad \text{and} \quad S^D = \{1, \ldots, K\}. \tag{11}$$

Accordingly, the vectors have the form:

$$\mathbf{p}^E(t) = (p_0(t), 0, \ldots, 0),$$

$$\mathbf{p}^D(t) = (0, p_1(t), \ldots, p_K(t)), \quad \mathbf{p}^D(t, x) = (0, p_1(t, x), \ldots, p_K(t, x)). \tag{12}$$

And the matrices are given by:

$$\mathbf{Q}^{E,E} = \begin{bmatrix} -\lambda & 0 & \cdots & 0 \\ 0 & \cdots & & 0 \\ \vdots & & & \vdots \\ 0 & \cdots & & 0 \end{bmatrix}, \quad \mathbf{Q}^{E,D} = \begin{bmatrix} 0 & \lambda & \cdots & 0 \\ 0 & \cdots & & 0 \\ \vdots & & & \vdots \\ 0 & \cdots & & 0 \end{bmatrix}, \tag{13}$$

$$\mathbf{Q}^D = \begin{bmatrix} 0 & \cdots & & & 0 \\ 0 & -\lambda & \lambda & \cdots & 0 \\ \vdots & \ddots & \ddots & \ddots & \vdots \\ \vdots & & \ddots & -\lambda & \lambda \\ 0 & \cdots & & 0 & 0 \end{bmatrix}$$

$$\Delta E = \begin{bmatrix} 0 & \cdots\cdots\cdots & 0 \\ 1 & 0 & \cdots\cdots & 0 \\ 0 & \cdots\cdots\cdots & 0 \\ 0 & \cdots\cdots\cdots & 0 \\ 0 & \cdots\cdots\cdots & 0 \end{bmatrix}, \quad \Delta D = \begin{bmatrix} 0 & \cdots\cdots\cdots & 0 \\ 0 & \cdots\cdots\cdots & 0 \\ 0 & 1 & 0 & \cdots & 0 \\ \vdots & \ddots & \ddots & \ddots & \vdots \\ 0 & \cdots & 0 & 1 & 0 \end{bmatrix}. \qquad (14)$$

We assume that the system is initially in state $n = 0$. Single state equations are yielded by inserting the vectors and matrices into the matrix state equations (6), (7), (8), (9), and (10). For state $n = 0$ the following ODE and initial condition is obtained:

$$\frac{d}{dt} p_0(t) = -\lambda p_0(t) + p_1(t, \tau), \qquad (15)$$
$$p_0(0) = 1. \qquad (16)$$

For state $n = 1$ the following PDE, boundary, and initial condition is obtained:

$$\left(\frac{\partial}{\partial t} + \frac{\partial}{\partial x} \right) p_1(t, x) = -\lambda p_1(t, x), \quad \text{for } 0 < x \leq \tau, \qquad (17)$$
$$p_1(t, 0) = \lambda p_0(t) + p_2(t, \tau), \qquad (18)$$
$$p_1(0, x) = 0, \quad \text{for } x \geq 0. \qquad (19)$$

For states $1 < n < K$ the following PDE, boundary, and initial condition is yielded:

$$\left(\frac{\partial}{\partial t} + \frac{\partial}{\partial x} \right) p_n(t, x) = -\lambda p_n(t, x) + \lambda p_{n-1}(t, x), \quad \text{for } 0 < x \leq \tau, \qquad (20)$$
$$p_n(t, 0) = p_{n+1}(t, \tau), \qquad (21)$$
$$p_n(0, x) = 0, \quad \text{for } x \geq 0. \qquad (22)$$

And for the state $n = K$ the following three equations are yielded:

$$\left(\frac{\partial}{\partial t} + \frac{\partial}{\partial x} \right) p_K(t, x) = \lambda p_{K-1}(t, x), \quad \text{for } 0 < x \leq \tau, \qquad (23)$$
$$p_K(t, 0) = 0, \qquad (24)$$
$$p_K(0, x) = 0, \quad \text{for } x \geq 0. \qquad (25)$$

Finally, the integral equations for states $1 \leq n \leq K$ are given by:

$$p_n(t) = \int_0^\tau p_n(t, x) dx. \qquad (26)$$

Equations (15) – (26) are the single state equations describing the transient behavior of the system.

3 Solution of the State Equations

The solution of the state equations introduced in Section 2 is not trivial, because they consist of partial and ordinary differential equations which are coupled by complex boundary conditions. However, the boundary conditions are less complex than in the case of the steady-state solution of DSPNs with two concurrently enabled deterministic transitions [12]. It turns out that the system of PDEs can be solved in isolation. It is then possible to iteratively solve the system of ODEs using standard Runge-Kutta methods [21]. In the special case that in each marking a deterministic transition is enabled (i.e. $\mathcal{S}^E = \emptyset$), the system of ODEs is empty and the boundary condititions are less complex. In this special case a simple and efficient solution is possible. Understanding the special case is also helpful for understanding the more general case. In Subsection 3.1 the system of PDEs is solved. A solution technique for the special case is derived in Subsection 3.2 and for the general case in Subsection 3.3. In Subsection 3.4 we propose to use the solution technique for the special case as an approximation method in more general cases.

3.1 Solution of the PDEs

Equation (6) represents a system of partial differential equations of first order and with constant coefficients. The matrix equation is very similar to the matrix equation which describes the transient behavior of a CTMC. For the solution one may use the property that both the transient time t and the age variable x increase with the same speed. It is therefore possible to reduce the values of $\mathbf{p}^D(t, x)$ to the values on the boundary. Note that we are only interested in values in the upper right quadrant, i.e. $x, t \in \mathbb{R}^+$. Arbitrary values of $\mathbf{p}^D(t, x)$ depend therefore on the values of $\mathbf{p}^D(t, x)$ on the positive parts of the axes. Using either the method of characteristics [11], which results in a variable substitution, or the double Laplace transform the following solution of (6) can be derived [12]:

$$\mathbf{p}^D(t, x) = \begin{cases} \mathbf{p}^D(t-x, 0) \cdot e^{\mathbf{Q}^D \cdot x} & \text{for } x \leq t \\ \mathbf{p}^D(0, x-t) \cdot e^{\mathbf{Q}^D \cdot t} & \text{for } x > t \end{cases} \quad (27)$$

The solution is visualized in Fig. 2. The main diagonal divides the upper right quadrant into two parts. Values in the upper part depend on values of the x-axis and values in the lower part depend on the boundary values of the t-axis. In both cases the boundary values have to be multiplied with a matrix exponential. The solution of $\mathbf{p}^D(t, x)$ is therefore reduced to the solution of the values on the axes given by $\mathbf{p}^D(0, x-t)$ or $\mathbf{p}^D(t-x, 0)$.

3.2 Special Case: Vanishing System of ODEs

Now the special case is considered that in each marking a deterministic transition is enabled. In that case $\mathcal{S}^E$ is empty and therefore $\mathbf{p}^E(t)$ is given by the null

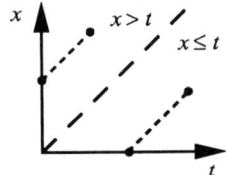

Fig. 2. Dependence of the values on the boundary values

vector. As a consequence, the system of ODEs (7) reduces to the empty system $0 = 0$ and the boundary conditions (9) reduce to:

$$\mathbf{p}^D(t, 0) = \mathbf{p}^D(t, \tau) \cdot \Delta^D. \tag{28}$$

An easy to implement solution formula can be derived for that less complex system of state equations. For notational convenience, we introduce l and s as:

$$l = \left\lfloor \frac{t}{\tau} \right\rfloor, \quad s = t - l \cdot \tau. \tag{29}$$

For a given transient instant of time t, l denotes how often a deterministic transition has fired and s denotes the elapsed time since the last firing. In Fig. 3 a scheme is shown how the stochastic process evolves. The probability mass of the densities $\mathbf{p}^D(t, x)$ is concentrated at the points $\left(t, t - \left\lfloor \frac{t}{\tau} \right\rfloor \cdot \tau\right) = (t, s)$. These points are shown as solid lines in the figure. The values at all other points are equal to zero. $\mathbf{p}^D(t, x)$ is therefore a vector of impulses and can be written as

$$\mathbf{p}^D(t, x) = \mathbf{q}(t, x) \cdot \delta(x - s), \tag{30}$$

where $\mathbf{q}(t, x)$ denotes the area and the Dirac delta function $\delta(x - s)$ gives the location of the impulses.

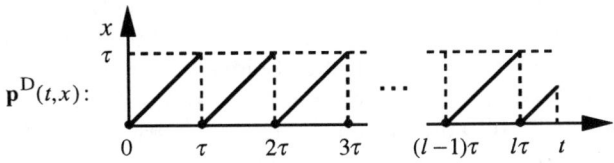

Fig. 3. Scheme for a vanishing system of ODEs

Using the integral equation (10) and the sifting property of the Dirac delta function leads to:

$$\mathbf{p}^D(t) = \int_0^T \mathbf{p}^D(t,x) dx = \int_0^T \mathbf{q}(t,x) \cdot \delta(t-s) dx = \mathbf{q}(t,s). \tag{31}$$

The probabilities in t are thus given by the area of the impulses in point (t,s). In the following expressions are obtained which relate the impulses in (t,s) to the initial impulses in $(0,0)$. The solution formula for PDEs (27) leads to:

$$\mathbf{p}^D(t,s) = \mathbf{p}^D(t-s,0) \cdot e^{\mathbf{Q}^D \cdot s} = \mathbf{p}^D(l \cdot \tau, 0) \cdot e^{\mathbf{Q}^D \cdot s}. \tag{32}$$

Employing the simplified boundary equation (28) yields:

$$\mathbf{p}^D(l \cdot \tau, 0) = \mathbf{p}^D(l \cdot \tau, \tau) \cdot \Delta^D. \tag{33}$$

Using again the solution formula for the PDEs (27) we obtain:

$$\mathbf{p}^D(l \cdot \tau, \tau) = \mathbf{p}^D((l-1) \cdot \tau, 0) \cdot e^{\mathbf{Q}^D \cdot \tau}. \tag{34}$$

The initial impulses are given by:

$$\mathbf{p}^D(0,0) = \mathbf{p}^D(0) \cdot \delta(x). \tag{35}$$

Employing Eqns. (32) – (35) recursively in order to compute the area of the impulses $\mathbf{q}(t,x)$ in Eq. (31) leads to the following single solution formula for the special case considered in this section:

$$\mathbf{p}^D(t) = \mathbf{p}^D(0) \cdot \prod_{i=1}^{l} \left(e^{\mathbf{Q}^D \cdot \tau} \cdot \Delta^D \right) \cdot e^{\mathbf{Q}^D \cdot s}. \tag{36}$$

The vector of transient probabilities is given by multiplying the vector of initial probabilities l times with the matrix exponential in τ and with Δ^D and by subsequently multiplying it with the matrix exponential in s. The transient solution is thus reduced to matrix multiplications and to the transient solution of a CTMC for which efficient numerical techniques are well-known. This solution is similar to the solution of phased-mission systems [4] where each phase is represented by a CTMC and the phase durations are constant. Note that the solution formula (36) could be extended for cases in which a fixed firing sequence of deterministic transitions can be identified. The DSPN model of a fault-tolerant clocking system described in [19] is an example for which that is possible. The result of this section generalizes the results which were obtained in [16].

3.3 General Case

In case the set $\mathcal{S}^E$ is not empty, the ODEs do not vanish and the solution is therefore more difficult, but it is still possible to iteratively determine the transient probabilities. Figure 4 shows a scheme for the solution. Since $\mathbf{p}^E(t)$ depends only on t, its definition set is one-dimensional and can be divided into intervals of length τ. $\mathbf{p}^D(t,x)$ depends both on t and x subject to the constraint that $x \leq \tau$, therefore the definition set is two-dimensional and can be divided into squares of length τ. The initial conditions determine the values at $\mathbf{p}^E(0)$ and $\mathbf{p}^D(0,x)$. The transient probabilities can be obtained by iteratively stepping over the intervals and squares. Due to the solution of the PDEs (27) the values of $\mathbf{p}^D(t,\tau)$ at the upper boundary of the squares are determined by the values $\mathbf{p}^D(t-\tau,0)$. The values of $\mathbf{p}^E(t)$ are then determined by the ODEs (7). The values of $\mathbf{p}^D(t,0)$ are related to the values of $\mathbf{p}^D(t,\tau)$ via the boundary conditions (9). In each step the values of $\mathbf{p}^E(t)$ and of $\mathbf{p}^D(t,0)$ are determined. Since these are functions in t, a discretization is necessary. For the solution of the ODEs, standard Runge-Kutta methods can be used.

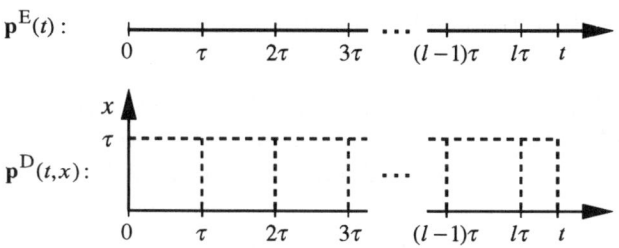

Fig. 4. Scheme for the general case

Step 1. The values of $\mathbf{p}^E(t)$ and $\mathbf{p}^D(t,0)$ in the first interval (i.e. for $0 \leq t < \tau$) are determined. The initial condition determines the values in $t = 0$:

$$\mathbf{p}^E(0) \text{ given, } \mathbf{p}^D(0) \text{ given, and } \mathbf{p}^D(0,x) = \mathbf{0}, \text{ for } x > 0. \tag{37}$$

Employing Eq. (27) shows that the values at the upper boundary of the first square are equal to zero:

$$\mathbf{p}^D(t,\tau) = \mathbf{p}^D(0,\tau - t) \cdot e^{\mathbf{Q}^D \cdot t} = \mathbf{0}. \tag{38}$$

Hence the ODEs (7) do not contain the term $\mathbf{p}^D(t,\tau) \cdot \mathbf{\Delta}^E$ in that interval. The solution for $\mathbf{p}^E(t)$ is therefore given by:

$$\mathbf{p}^E(t) = \mathbf{p}^E(0) \cdot e^{\mathbf{Q}^{E,E} \cdot t}. \tag{39}$$

Subsequently using the boundary conditions (9) leads to a solution for $\mathbf{p}^D(t,0)$:

$$\mathbf{p}^D(t,0) = \mathbf{p}^E(t) \cdot \mathbf{Q}^{E,D}. \tag{40}$$

The results of this step have to be stored at discrete points. In a first approach a fixed stepsize is used. Better results could be obtained with an adaptive stepsize.

Step $l \to l+1$. Given that the values of $\mathbf{p}^E(t)$ and $\mathbf{p}^D(t,0)$ are known for the l-th interval (i.e. $(l-1) \cdot \tau \leq t < l \cdot \tau$) the values of these vectors are determined for the $(l+1)$-th interval (i.e. $l \cdot \tau \leq t < (l+1) \cdot \tau$). The values at the upper boundary of the $(l+1)$-th square are given by the solution of the PDEs (27):

$$\mathbf{p}^D(t,\tau) = \mathbf{p}^D(t-\tau,0) \cdot e^{\mathbf{Q}^D \cdot \tau}. \tag{41}$$

Having obtained the values of $\mathbf{p}^D(t,\tau)$, the values of $\mathbf{p}^E(t)$ can be determined from the system of ODEs (7). Numerical standard algorithms like Runge-Kutta methods [21] can be used for the numerical computation. In a first approach we choose the explicit Runge-Kutta method of fourth order. Subsequently, the boundary conditions (9) can be employed in order to yield the values $\mathbf{p}^D(t,0)$ at discrete points of the lower boundary of the square. Note that the number of ODEs corresponds to the number of states in $\mathcal{S}^E$.

Integration Step. After the computation of the values of $\mathbf{p}^D(t,0)$ the transient state probabilities can be computed by numerical integration. Formally setting $\mathbf{p}^D(t,0) = 0$ for $t < 0$ the vector of transient state probabilities in states of $\mathcal{S}^D$ is given by:

$$\mathbf{p}^D(t) = \int_0^\tau \mathbf{p}^D(t-x,0) \cdot e^{\mathbf{Q}^D \cdot x} dx. \tag{42}$$

The trapezoid rule can be used for the numerical computation.

3.4 Approximate Solution of the State Equations

Since the solution formula derived for the special case in Section 3.2 is very simple, it seems also useful to employ it for the approximate solution in more general cases. If the condition $\mathcal{S}^E = \emptyset$ is not satisfied, a sufficiently small artificial time step can be introduced. All deterministic transitions are represented as multiples of that time step. Furthermore, it can be enforced that the deterministic transitions may become enabled only at multiples of the time step. This leads to an approximation of the underlying stochastic process. However, it seems reasonable to use appropriate approximations in the solution process, since even the model is an approximation.

Employing the proposed assumptions, a DSPN without structural restrictions can be approximately analyzed by means of Eq. (36). The introduced artificial timestep leads to a discretization and thus to a state space expansion. The costs of the approximate method are given by the costs for computing the matrix exponential and for performing the matrix multiplications. If the timestep becomes small, the system is less stiff and computing the matrix exponential becomes less expensive. The costs are then mainly determined by the number of matrix multiplications.

4 Numerical Examples

Numerical results are presented which are obtained by a Mathematica prototype implementation of the algorithms.

4.1 Machine-Repairman System

We consider a DSPN model of a machine-repairman system. The system consists of K machines which operate in the daytime and which are attended at night. The machines may fail during operation. A crew repairs the failed machines at night. Figure 5 shows the DSPN. The DSPN is a simplified version of the model of a fault-tolerant clocking system described in [19].

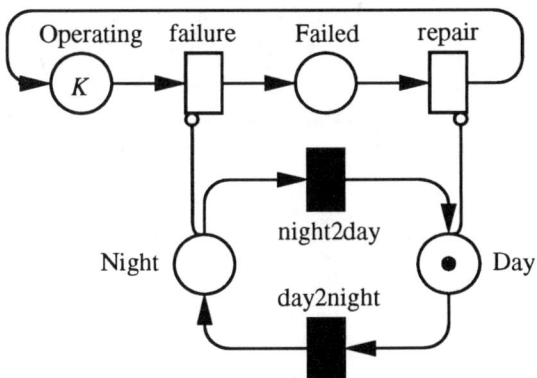

Fig. 5. DSPN model of a machine-repairman system

The upper subnet of the DSPN models the failure (exponential transition *failure* with marking-dependent rate: product of number of tokens in place *Operating* and of λ), and repair (exponential transition *repair* with rate: μ). The lower subnet models the change between day and night (deterministic transition *day2night* with delay: τ) and between night and day (deterministic transition *night2day* with delay: τ). The inhibitor arcs cause that *failure* is not enabled when a token is in place *Night* and *repair* is not enabled when a token is in place *Day*.

In the DSPN in each marking a deterministic transition is enabled with the same firing time. Therefore the special solution formula derived in Section 3.2 can be used for the transient analysis of the DSPN. In the following experiments the parameters are chosen according to:

$$K = 3, \ \tau = \frac{1}{2}, \ \lambda = \frac{1}{7}, \ \mu = 1.$$

Figure 6 shows the expected number of operating machines (corr. to the mean value of tokens in *Operating*) over a period of 10 days and Figure 7 shows the probability that all machines have failed (corr. to the probability that no token is in *Operating*) over the same time period. A logarithmic scale is used for the second curve.

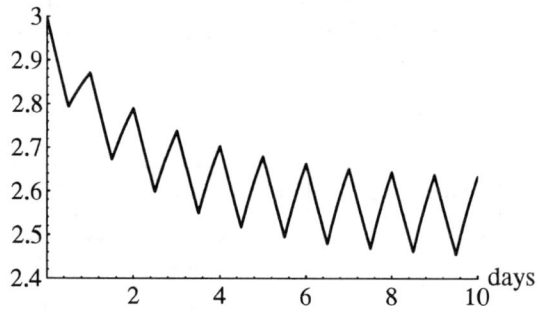

Fig. 6. Expected number of operating machines

4.2 M/D/1/K Queueing System

We consider an M/D/1/K queueing system as described in Sec. 2.3. In the following experiments the parameters of the queueing system are chosen as:

$$K = 3, \lambda = 4, \tau = 1.$$

Figure 8 shows the probabilitiy $p_0(t)$ that no customers are inside the system for values $0 \leq t \leq 10$. The horizontal line corresponds to the stationary solution obtained with the method in [3]. The solid curve is obtained by the method proposed in Sec. 3.3 using explicit Runge-Kutta with a stepsize of 0.01. The dashed curve is obtained by the approximation method proposed in Sec. 3.4 with a time step of 0.1. The dot-dashed curve shows the values that are obtained if the deterministic delay is approximated by an Erlang distribution. The number

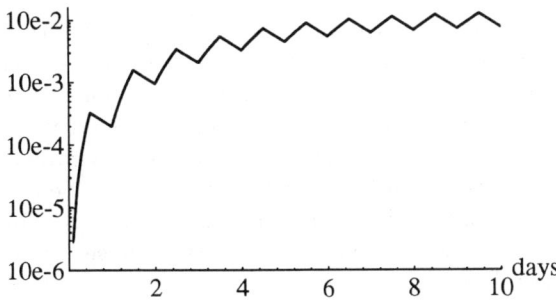

Fig. 7. Probability that all machines have failed

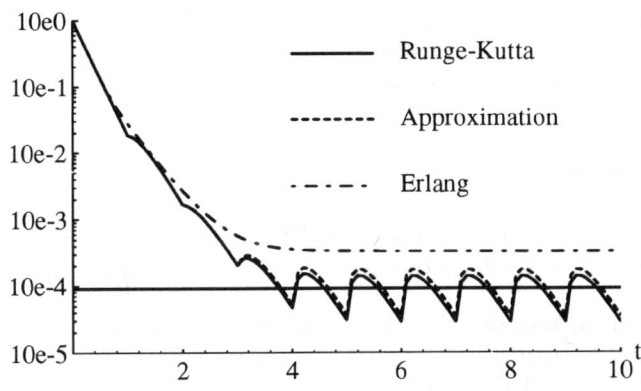

Fig. 8. Probability that no customer is inside the system

of phases is set to ten which corresponds roughly to the costs for the other approximation method.

The figure shows that the derivative of the function of the transient probability has jumps. Furthermore, in the considered case only time averages are existing, but no steady-state probability. The approximation according to Sec. 3.4 is well suited to achieve good results with an easy implementation and with low costs. Further experiments showed that the results converge against the results obtained with the Runge-Kutta method if the artificial timestep becomes smaller. An approximation using an Erlang distribution leads to worse results, which do not reflect the shape of the curve.

5 Conclusions

In this paper we investigated the transient analysis of deterministic and stochastic Petri nets (DSPNs) by the method of supplementary variables. State equations were derived which describe the transient behavior of a DSPN under certain structural conditions. If in all states of the DSPN a deterministic transition with the same firing delay is enabled, an efficient and easy to implement solution of the state equations is possible. In this case the solution can be based on matrix multiplications and on the transient solution of a continuous time Markov chain. In the more general case, an iterative scheme was proposed that uses Runge-Kutta standard methods. Furthermore, it was proposed to use the solution technique of the special case for the approximate transient analysis of DSPNs without structural restrictions. Numerical results for a model of a machine-repairman system and for a M/D/1/K queueing system were given, demonstrating the use of the different solution algorithms.

Acknowledgements

The author would like to thank Dimitris Logothetis for draft versions of his papers and for the comparison of numerical values.

References

1. M. Ajmone Marsan. Stochastic Petri Nets: An Elementary Introduction. *Advances in Petri Nets 1889, Lecture Notes in Computer Science 424*, pp. 1–29, Springer 1990.
2. M. Ajmone Marsan, G. Balbo, A. Bobbio, G. Chiola, G. Conte, A. Cumani. The Effect of Execution Policies on the Semantics of Stochastic Petri Nets. *IEEE Trans. Softw. Engin.*, **15** (1989) 832-846.
3. M. Ajmone Marsan, G. Chiola. On Petri Nets with Deterministic and Exponentially Distributed Firing Times. *Advances in Petri Nets 1986, Lecture Notes in Computer Science 266*, pp. 132–145, Springer 1987.
4. M. Alam, U.M. Al-Saggaf. Quantitative Reliability Evaluation of Repairable Phased-Mission Systems Using Markov Approach. *IEEE Trans. on Rel.*, **R-35** (1986) 498–503.
5. A. Bobbio, M. Telek. Transient Analysis of a Preemptive Resume M/D/1/2/2 through Petri Nets. Internal paper, Universit di Brescia, Italy, 1994, (submitted for publication).
6. G. Brat, M. Malek. Incorporating Delays and Exponential Distributions in Petri Nets for Responsive Systems Modeling. Internal paper, University of Texas, Austin, Texas, USA.
7. H. Choi, V. G. Kulkarni, K. S. Trivedi. Transient Analysis of Deterministic and Stochastic Petri Nets. *Proc. 14th Int. Conf. on Application and Theory of Petri Nets*, Chicago, IL, USA, pp. 166–185, 1993.
8. E. Cinlar. *Introduction to Stochastic Processes*. Prentice Hall, Englewood Cliffs, 1975.

9. G. Ciardo, R. German, C. Lindemann. A Characterization of the Stochastic Process Underlying a Stochastic Petri Net. *Proc. 5th Int. Workshop on Petri Nets and Performance Models (PNPM '93)*, Toulouse, France, pp. 170–179, 1993.
10. D.R. Cox. The Analysis of Non-Markov Stochastic Processes by the Inclusion of Supplementary Variables. *Proc. Camb. Phil. Soc. (Math. and Phys. Sciences)*, **51** (1955) 433–441.
11. R. Courant, D. Hilbert. *Methods of Mathematical Physics*, Vol. 2. Interscience, New York, 1962.
12. R. German, C. Lindemann. Analysis of Stochastic Petri Nets by the Method of Supplementary Variables. *Performance Evaluation*, **20** (1994) 317–335.
13. J. Keilson, A. Kooharian. On Time Dependent Queuing Processes. *Ann. Math. Stat.*, **31** (1960) 104–112.
14. D.G. Kendall. Stochastic Processes Occuring in the Theory of Queues and their Analysis by the Method of the Imbedded Markov Chain. *Ann. Math. Stat.*, **24** (1953) 338–354.
15. V. G. Kulkarni. *Lecture Notes on Stochastic Models in Operation Research*. University of North Carolina, Chapel Hill, NC, USA, 1990.
16. D. Logothetis, K. S. Trivedi. Transient Analysis of the Leaky Bucket Rate Control Scheme Under Poisson and ON-OFF Sources. *Proc. of INFOCOM '94*, Toronto, Canada, 1994.
17. D. Logothetis, K. S. Trivedi. Time-Dependent Behavior of Redundant Systems with a Single Repairperson and Deterministic Repair. Internal paper, Duke University, Durham, NC, USA, 1994.
18. D. Logothetis, K. S. Trivedi. The Effect of Detection and Restoration Times for Error Recovery in Communication Networks. Internal paper, Duke University, Durham, NC, USA, 1994.
19. M. Lu, D. Zhang, T. Murata. Analysis of Self-Stabilizing Clock Synchronization by Means of Stochastic Petri Nets. *IEEE Trans. on Comp.*, **39** (1990) 597–604.
20. M. Malek. Responsive Systems: A Challenge for the Nineties. *Proc. of EUROMICRO '90*, pp. 9–16, 1990.
21. J. Stoer, R. Burlirsch. *Introduction to Numerical Analysis*. Springer-Verlag, 1980.

Discrete Time Deterministic and Stochastic Petri Nets

Robert Zijal *

Technische Universität Berlin, Institut für Technische Informatik,
Fachgebiet Prozeßdatenverarbeitung und Robotik,
(Real-Time Systems and Robotics),
Franklinstr. 28/29, 10587 Berlin, F.R.Germany,
e-mail: bob@cs.tu-berlin.de

Abstract. This paper presents a new approach to stochastic Petri nets (SPNs) with a discrete time scale. SPNs are considered in which the timed transitions may fire after a constant delay (deterministic transitions) or after a geometrically distributed delay (geometric transitions). Since a discrete time scale is used, deterministic and geometric transitions can be mixed without structural restrictions. Therefore this approach overcomes the drawback of the class of deterministic and stochastic Petri nets (DSPNs), in which at most one deterministic transition may be enabled in each marking of the net. A classification of possible events in each state is introduced in order to automatically resolve all conflicts between transitions and to map the SPN to an underlying discrete time Markov chain. It is thus possible to compute the transient and steady state solution of the net with standard techniques.

1 Introduction

Several classes of *stochastic Petri nets* (SPNs) have been proposed in order to define a unified framework for the modeling and analysis of concurrent systems. Qualitative properties such as liveness or absence of deadlocks can be studied by structural properties (see e.g. [Mur89]). Most commonly, in a SPN the ordinary Petri net is augmented by continuous time random variables which specify the transition firing times. As a consequence, a stochastic process is underlying the SPN and can be studied in order to obtain quantitative measures of the model. In [ABC84] the class of *generalized stochastic Petri nets* was defined where the transitions fire either without delay (immediate transitions) or after an exponentially distributed delay. The underlying stochastic process is a continuous time Markov chain and the transient and steady-state analysis can be obtained with standard techniques (see e.g. [GH85, Cia&al93]).

---

* Robert Zijal was supported by a doctoral fellowship from the German National Research Council under grant Ho 1257/1-2.

Our main objective is to model technical sytems where the duration of most activities is constant (e.g. transfer time in a communication system, repair time in a fault-tolerant system) and where the duration of some activities is modeled by a probability distribution (e.g. randomly occuring failures). The class of *deterministic and stochastic Petri nets* (DSPNs) [AC87] contains transitions which fire after an exponentially distributed or deterministic delay. Under the structural restriction that at most one transition with a deterministic delay is enabled in each marking the underlying stochastic process is a semi-regenerative process [Çin75] and formulas for the steady-state solution were derived in [AC87] considering an embedded Markov chain. In order to tackle the structural restriction it was shown in [GL93], that state equations describing the dynamic behaviour of DSPNs without that restriction can be derived by the method of supplementary variables. The state equations contain partial differential equations and are difficult to solve.

Apart from continuous time also discrete time has been considered. Molloy developed the class of *discrete time stochastic Petri nets* (DTPNs) in [Mol85], where the transition firing times are specified by a geometric distribution with the same time step. Holliday and Vernon introduced *generalized timed Petri nets* (GTPNs) in [HV87]. In both models the underlying stochastic process is a *discrete time Markov chain* (DTMC). Since a deterministic firing time is a special case of a geometrically distributed firing time, a deterministic transition can be represented in a SPN with discrete time. In order to represent distributions with different time steps a discretization of the time step can be used. In a DTPN the user has to model the discretization with net constructs (a series of places and transitions). As a consequence, conflicts have to be resolved manually by the user. In a GTPN geometric distributions have to be represented by net constructs consisting of deterministic transitions and a looping arc, which becomes more complicated in case of conflicts. In [CGL93] it was outlined that an expansion of the state space can be used to allow geometric firing time distributions with different time steps.

In this paper we define the class of *discrete time deterministic and stochastic Petri nets* (dtDSPNs). In a dtDSPN transitions fire either without time (immediate transitions) or after a geometrically distributed time. Transitions with a deterministic firing delay are a special case. Arbitrary time steps are allowed. All types of transitions can be mixed *without* structural restrictions. Moreover, since a discrete time scale is used, simultaneous firings of transitions in the same instant of time are possible. An algorithm is presented which resolves all conflicts automatically and computes the one-step transition probability matrix of the underlying DTMC. The transient and stationary solution can then be obtained by standard techniques (see e.g. [GH85, Cia&al93]). Although the state space may become large in case the greatest common divisor of the time steps of all firing time distributions is small, the approach is well suited to model technical systems where the most activities have a constant duration and failures can be modeled by random variables.

The rest of the paper is organized as follows. In Section 2 the considered class of dtDSPNs is introduced. In Section 3 a general description of the solution technique is presented and in Section 4 numerical examples are given. Concluding remarks are given in Section 5.

2 The Considered Class of Stochastic Petri Nets

A dtDSPN is formally given by a nine-tuple
$$\text{dtDSPN} = (P, T, I, O, H, \mathcal{P}, \mathbf{m_0}, F, W).$$

P is the set of places. Places may contain tokens. The vector $\mathbf{m} \in \mathbb{N}_0^{|P|}$ of number of tokens in each place is referred to as *marking*.

T is the set of transitions and can be partitioned into the set T_{timed} of timed transitions firing after a certain delay and into the set T_{imm} of immediate transitions firing without consuming time.

I, O, and H represent the input, output, and inhibitor arcs of the transitions, respectively. The arc multiplicities may be marking-dependent:
$I : P \times T \times \mathbb{N}_0^{|P|} \rightarrow \mathbb{N}_0$, $O : T \times P \times \mathbb{N}_0^{|P|} \rightarrow \mathbb{N}_0$, $H : P \times T \times \mathbb{N}_0^{|P|} \rightarrow \mathbb{N}_0$.
Marking-dependent arc multiplicities increase the modelling power. The firing of one transition for example can flush an arbitrary number of tokens from a place.

$\mathcal{P} : T \rightarrow \mathbb{N}_0$ assigns a priority to each transition. Timed transitions have lowest priority equal to zero and immediate transitions have priorities greater than zero. $\mathbf{m_0}$ is the initial marking of the net.

The usual firing rules are assumed. The places connected by input (output) arcs to a transition t are referred to as input (output) places of t. t is *enabled* if all of its input places contain at least the number of tokens equal to the multiplicity of the input arc and all places connected by inhibitor arcs contain a number of tokens less than the multiplicity of the inhibitor arcs:
$$\forall p \in P : I(p, t, \mathbf{m}) \leq \#(p, \mathbf{m}) \land (H(p, t, \mathbf{m}) > \#(p, \mathbf{m}) \lor H(p, t, \mathbf{m}) = 0),$$
where $\#(p, \mathbf{m})$ means the number of tokens in place p in $\mathbf{m}$. Additionally, no transition with a higher priority may fulfill this condition.

An enabled transition may *fire* by removing tokens from the input places and by adding tokens to the output places according to the multiplicities of the arcs.

All markings reachable from the initial marking constitute the *reachability set* of the net. The *reduced reachability set* is given by all *tangible markings* enabling only timed transitions.

$F : T_{timed} \rightarrow \mathcal{F}$ assigns a geometric firing time distribution to each timed transition. $\mathcal{F}$ is the set of all geometric distributions, denoted by $Geom(p, n\omega)$, given by the probability p ($p \in (0, 1]$) and the time step $n \cdot \omega$ ($n \in \mathbb{N}$, $\omega \in (0, \infty)$). n may be arbitrarily chosen for each transition, wheras ω has to be kept fixed. ω is the greatest common divisor of the time steps of all firing time distributions and therefore the underlying time step of the model. Transitions fire with probability p or they fire not with probability $(1-p)$ only at time instances which are multiples of ω. The probability mass function of the firing time X (Fig.: 1) is given by
$$Pr\{X = i \cdot n \cdot \omega\} = p \cdot (1-p)^{i-1}, \qquad i \in \mathbb{N}$$
and the mean is given by: $\frac{n\omega}{p}$.

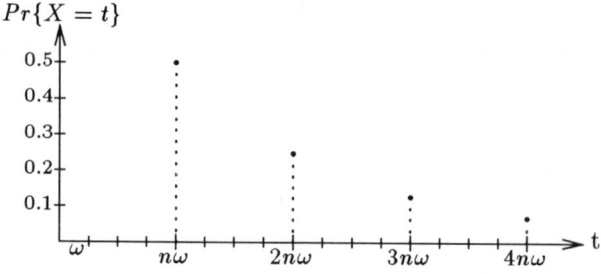

Fig. 1. Probability mass function of $Geom(0.5, 4\omega)$.

A deterministic firing time, denoted by $Const(n\omega)$, is a special case of a geometric distribution (Fig.: 2): setting $p = 1$ leads to $Const(n\omega) = Geom(1, n\omega)$.

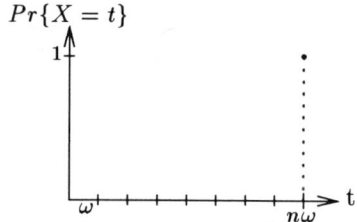

Fig. 2. Probability mass function of $Const(8\omega)$.

In a dtDSPN, transitions with deterministic and geometrically distributed firing time can be mixed *without* structural restrictions. The use of a discrete time scale allows the transitions to fire in the same instant of time. Since this may lead to conflicts, weights are needed.

$W : T \rightarrow (0, \infty)$ assigns a weight to each transition. The default weight is one. Weights are used to resolve conflicts between transitions.

3 Description of the Solution Technique

The stochastic process underlying a dtDSPN is a DTMC. In Section 3.1 the state space of the DTMC is formally defined and in Section 3.2 a method is presented for the generation of the state space and for the computation of the one-step transition probabilities of the DTMC. In Section 3.3 the role of the deterministic firing times in the solution process is discussed. The calculation of measures of interest is presented in Section 3.4 and some state reduction techniques in Section 3.5.

3.1 Definition of the DTMC

The time step of the DTMC is given by ω. Since the time steps of the firing time distributions may be multiples of ω, the state space of the DTMC is represented by the tangible markings of the net and by the vector of *remaining firing times* (RFTs) of the transitions.

A RFT $r > 0$ of a transition represents that the time step of the transition is reached after the time $r \cdot \omega$ has elapsed. For a disabled transition the RFT is set to zero.

Let the set of all tangible markings of the net be denoted as $\mathcal{M} \subset \mathbb{N}_0^{|P|}$, and let the set of all possible vectors of RFTs be denoted as $\mathcal{R} \subset \mathbb{N}_0^{|T_{timed}|}$.

The set of states of the DTMC is then denoted as:
$$\mathcal{S} \subseteq \mathcal{M} \times \mathcal{R}.$$
We consider only the case of a finite state space.

3.2 Construction of the DTMC

The states and the one-step transition probabilities of the DTMC can be determined by starting with the initial state and by visiting all reachable states.

In the following we assume that the initial marking is tangible [1]. The initial state is given by $s_0 = (\mathbf{m_0}, \mathbf{r_0})$. The entries of $\mathbf{r_0}$ corresponding to enabled transitions are set to their predefined RFT and to zero otherwise.

For each state $s = (\mathbf{m}, \mathbf{r})$ the set $\mathcal{S}'$ of states reachable at the next time step can be determined. The states of $\mathcal{S}'$ and the corresponding one-step transition probabilities can be computed by the following four steps:

Step 1:
Determine the set of the *firing enabled* timed transitions. These transitions may fire after the next time step. The set T_{fe} in state $s = (\mathbf{m}, \mathbf{r})$ is given by
$$T_{fe} = \{t \in T_{timed} \mid t \text{ enabled in } \mathbf{m}, \text{ RFT of } t \text{ equal to } 1\},$$
where a RFT equal to 1 means, that t will be able to fire after the next discrete time step ω.

Step 2:
Simultaneous firings of transitions from T_{fe} may occur on a discrete time scale at the next time step and cause conflicts. For this reason the transitions of T_{fe} are partitioned into *independent groups* of conflicting transitions. For each group *firing events* are identified. The probabilities of the firing events can be computed for each group in isolation.

First, a binary relation *direct conflict* $\mathcal{DC} \subset T_{fe} \times T_{fe}$ between transitions is defined. $\mathcal{DC}$ represents the competition of transitions for the tokens in the input places. Two transitions $t_i, t_j \in T_{fe}$ are in direct conflict in marking $\mathbf{m}$, if they

---
[1] Otherwise the set of tangible states reachable from the initial marking has to be considered.

share at least one common input place:
$$t_i \, \mathcal{DC} \, t_j \quad \Leftrightarrow \quad \exists p \in P : I(p, t_i, \mathbf{m}) > 0 \wedge I(p, t_j, \mathbf{m}) > 0.$$

Note that this definition differs from the ordinary definition of conflicts in Petri nets and is sufficient for the following two reasons. Inhibitor arcs have no influence, since only enabled transitions are considered. A discrete time scale is used, which forces timed transitions to fire only at discrete points in time. Different firing sequences for the same instant of time, which could lead to non-deterministic *confusions* (as for immediate transitions), do not exist.

Enabled transitions are in direct conflict even if their common input place contains more tokens than required for their firing. This is because in our model transitions with the same input place compete for the same token(s), which can be moved by the firing of only one transition at a time at the next time step. Therefore a situation is not possible, where directly conflicting transitions would fire simultaneously, due to a prior partition of tokens amongst them from their common input place. This convention simplifies the analysis, since the token distribution over transitions, which results in more possible firing combinations, needs not to be taken into account.

$\mathcal{DC}$ can be determined very easily. Performing the reflexive, symmetric, and transitive closure of $\mathcal{DC}$ leads to the equivalence relation

$$\text{\textit{indirect conflict}:} \quad \mathcal{IC} = \mathcal{DC}^*.$$

The equivalence classes of the quotient set $T_{fe}/_{\mathcal{IC}}$ are referred to as *independent groups* and represent sets of transitions for which the firing of transitions is not independent.

In the following we distinguish between three group types:
$$1 : n \text{ -}, \, m : 1 \text{ -, and } m : n \text{ - groups.}$$
For each group type the set of possible exclusive *firing events* is identified, and it is shown how the probabilities of the events can be computed.

$1 : n$ - group:
The group consists of one single transition t connected with n input places (Fig.: 3). Two events are possible: Either t fires or not. Let p denote the probability that t will fire in isolation. t fires with probability p and t does not fire with probability $1 - p$.

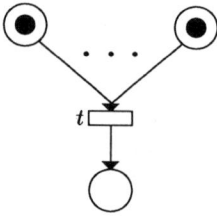

Fig. 3. One enabled transition forms a $1 : n$ - group.

$m : 1$ - group:

The group consists of m transitions with at least one common input place. $m+1$ events are possible. Since all transitions compete for the same token(s) on their common input place, either one of the m transition is allowed to fire or no transition fires.

As more than one transition may attempt to fire at the same instance of time, a conflict situation arises, that is resolved by using weights w_i for every transition t_i. The weights are by default equal to one, but can be changed in the Petri net model, where it seems appropriate. Possible firing combinations with their weighted probabilities determine the probabilities of the events, which can be computed for the most general case as follows.

Let the transitions of the group be enumerated by t_i, $i = 1, ..., m$. Furthermore, let p_i and $(1 - p_i)$ be the probabilities that transition t_i will fire or not fire in isolation, respectively. Define the index sets $I = \{1, ..., m\}$, $I_i = I \setminus \{i\}$. The probability P_{t_i} of the event that t_i fires, is given by:

$$P_{t_i} = p_i \cdot \sum_{I'_i \subseteq I_i} \left(\prod_{j \in I'_i} p_j \cdot \prod_{j \in I_i \setminus I'_i} (1 - p_j) \cdot \frac{w_i}{w_i + \sum_{j \in I'_i} w_j} \right),$$

and the probability P_N of the event, that no transition fires is given by:

$$P_N = \prod_{j \in I} (1 - p_j)$$

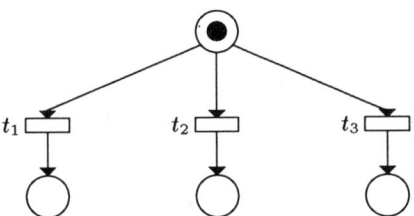

Fig. 4. Three enabled transitions in conflict, which form a $3 : 1$ - group.

The calculation of the four event probabilities for a $3 : 1$ - group (Fig.: 4) is exemplified in the following.

$P_{t_1} = p_1(1-p_2)(1-p_3) + p_1 p_2 (1-p_3)\frac{w_1}{w_1+w_2} + p_1(1-p_2)p_3\frac{w_1}{w_1+w_3} + p_1 p_2 p_3 \frac{w_1}{w_1+w_2+w_3}$

$P_{t_2} = p_2(1-p_1)(1-p_3) + p_2 p_1 (1-p_3)\frac{w_2}{w_1+w_2} + p_2(1-p_1)p_3\frac{w_2}{w_2+w_3} + p_1 p_2 p_3 \frac{w_2}{w_1+w_2+w_3}$

$P_{t_3} = p_3(1-p_1)(1-p_2) + p_3 p_1 (1-p_2)\frac{w_3}{w_1+w_3} + p_3(1-p_1)p_2\frac{w_3}{w_2+w_3} + p_1 p_2 p_3 \frac{w_3}{w_1+w_2+w_3}$

$P_N = \prod_{i=1}^{|T_j|}(1 - p_i)$

$m : n$ - group:

The group consists of m transitions connected with n input places, additionally there is no common input place for all transitions. In a $m : n$ - group at least $m + 1$ events are possible. As in a $m : 1$ - group each single transition may fire or no transition may fire. Furthermore, the simultaneous firing of more than one transition, which are not in direct conflict may be possible. The computation of the event probabilities is demonstrated for an example of a 3 : 2 - group (Fig.: 5), where the additional event of the simultaneous firing of t_1 and t_3 is possible. Let again p_i denote the probability of firing of transition t_i in isolation and let w_i denote the weight of transition t_i, $i = 1, 2, 3$. Possible events are the single firing of t_1, t_2 or t_3, the simultaneous firing of t_1 and t_3, and that no transition fires. The event probabilities are denoted as: P_{t_1}, P_{t_2}, P_{t_3}, $P_{t_1 t_3}$ and P_N, respectively. They are given by:

$P_{t_1} = p_1(1-p_2)(1-p_3) + p_1 p_2 (1-p_3) \frac{w_1}{w_1+w_2}$

$P_{t_2} = p_2(1-p_1)(1-p_3) + p_2 p_1 (1-p_3) \frac{w_2}{w_1+w_2} + p_2(1-p_1)p_3 \frac{w_2}{w_2+w_3} + p_1 p_2 p_3 \frac{w_2}{w_1+w_2+w_3}$

$P_{t_3} = p_3(1-p_1)(1-p_2) + p_3(1-p_1)p_2 \frac{w_3}{w_2+w_3}$

$P_{t_1 t_3} = p_1 p_3 (1-p_2) + p_1 p_3 p_2 \frac{w_1+w_3}{w_1+w_2+w_3}$

$P_N = \prod_{i=1}^{3}(1-p_i)$

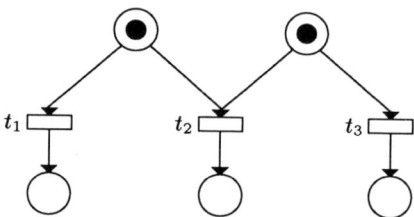

Fig. 5. Three enabled transitions in conflict, which form a 3 : 2 - group.

Step 3:
So far, independent groups of firing enabled transitions were built. For every group the firing events and their according probabilities were determined.

Let the groups be indexed by $k = 1, ..., n$, where $n = |T_{fe}/_{IC}|$. E^i denotes the set of – mutually exclusive – single events of group i.

Let the probabilities of the events of group i be indexed by $j = 1, ..., x$, where $x = |E^i|$. E^i_j denotes the event probability j of group i. All event probabilites of group i sum to one:

$$1 = \sum_{j=1}^{x} E^i_j$$

Events of independent groups take place simultaneously (one event per group). We refer to one combination of simultaneous events as *product event*. The set of all product events E is given by the cross product of all event sets:

$$E = E^1 \times \cdots \times E^n$$

The probability of a product event is given by the product of the single event probabilities: $\prod_{k=1}^{n} E^k_j$. All product event probabilities sum to one.

Step 4:
All product events are applied to the current DTMC state s.

After a product event has taken place in state $s = (\mathbf{m}, \mathbf{r})$, immediate transitions may be enabled. After the possible firing of immediate transitions a new state $s' = (\mathbf{m}', \mathbf{r}')$ is reached. $\mathbf{m}'$ is determined by the firing of the transitions. $\mathbf{r}'$ is determined as follows. Each entry of $\mathbf{r}$ greater than zero (the RFTs of enabled transitions) is decremented. After that each entry corresponding to a newly enabled transition in marking $\mathbf{m}'$ is set to the predefined RFT of the transition. And finally, the entries of transitions disabled in $\mathbf{m}'$ are set to zero.

The set $\mathcal{S}'$ of all reachable states from s is therefore given by the states reachable by product events combined with the possible firing of immediate transitions.

The one step transition probabilities of the DTMC are thus given by the product of the probabilities of the product events and of the probabilities of the immediate transition firing sequences. The probabilities of the immediate transition firing sequences can be determined by the algorithm presented in [Cia&al93].

3.3 The Geometric Distribution and Deterministic Firing Times

A deterministic firing time is a geometric distribution with probability $p = 1$. As a consequence, the probability of firing events involving transitions with a deterministic firing time may reduce to zero. The RFT is needed in order to discretize the firing time and to represent different deterministic firing times in the underlying DTMC. In case of a geometric distribution an arbitrary average firing time can be achieved for an arbitrary time step.

3.4 Extracting the Results from the DTMC

In Section 3.2 it was shown how the one-step transition probability matrix $\mathbf{P}$ of the DTMC can be constructed. The transient and stationary solutions of the DTMC can be obtained with standard techniques (e.g. [GH85, Cia&al93]). The probability distribution over states of the DTMC, $\pi = (\pi_1, ..., \pi_k)$, $k = |\mathcal{S}|$ is computed by the solution of the following linear equations:

$$\pi = \pi \mathbf{P}, \quad \sum_{i=1}^{k} \pi_i = 1.$$

Since $\mathbf{P}$ is usually a sparse matrix, sparse numerical techniques should be used.

The transient solution is defined as:

$$\pi^{(n)} = \pi^{(0)} \mathbf{P}^{(n)}, \text{ where } n \text{ is a time multiple of } \omega.$$

Let π_s denote the sojourn probability in a state $s = (\mathbf{m}, \mathbf{r})$, $s \in \mathcal{S}$. The probability distribution over markings can be then obtained by summing up all corresponding probabilities of the DTMC:

$$\pi_{\mathbf{m}} = \sum_{r \in \mathcal{R}} \pi_{(\mathbf{m},\mathbf{r})}, \quad \mathbf{m} \in \mathcal{M}.$$

The probability distribution of tokens in the places of the Petri net can be derived from the probability distribution over markings, which finally leads to the calculation of the mean number of tokens in a Petri net place.

The throughput of every transition t_i can be calculated automatically. Let p_i denote the firing probability and d_i the mean firing delay of t_i. The throughput $tput(t_i)$ is then given by:

$$tput(t_i) = \sum_f \pi_f \frac{p_i}{d_i}, \quad t_i \text{ may fire at the next time step } \omega \text{ in the state } s_f \in \mathcal{S}.$$

3.5 State Reduction

The size of the state space depends on the greatest common divisor ω of the time steps of all firing time distributions.

The geometric firing distribution contributes, compared to the constant distribution, more to the growth of the DTMC, due to its additional non-firing possibility when the firing time is reached.

In order to reduce states, a time step equal to ω should be used for a geometric distribution, whenever possible. The mean firing time is then given by $\frac{\omega}{p}$.

A better reduction of states can be reached by using an embedded Markov chain and by normalizing the results [Zub91, CGL93]. Therefore the solution method is slightly modified. The underlying stochastic process is then a discrete time semi Markov process.

Let T_e denote the set of enabled timed transitions and $rft(t)$ the RFT of transition t in state $s = (\mathbf{m}, \mathbf{r})$. The set of the firing enabled timed transitions in state s is then given by

$$T_{fe} = \{t_i \in T_e \mid \forall t_j \in T_e : t_i = min(rft(t_j))\}$$

The RFT entries of all enabled transitions are now reduced by the RFT of the firing transition, instead of ω, during the construction of the Markov chain. For this reason the RFT of the firing transition repesents the new holding time δ_s in state s of the now *embedded* DTMC (EDTMC). All i holding times are stored in the diagonal entry of a conversion matrix

$$\mathbf{C}, \text{ where } [c_{ij}] = \delta_i, i = j \text{ and } [c_{ij}] = 0, i \neq j$$

The steady-state solution of the EDTMC is computed by solving the linear system of equations:

$$\gamma = \gamma \mathbf{P}, \quad \sum_i \gamma_i = 1.$$

The embedded stationary probabilites in γ have to be *rescaled* using the holding times from conversion matrix $\mathbf{C}$:

$$\gamma' = \gamma \cdot \mathbf{C}$$

The stationary probability distribution π over states of the DTMC is obtained by the normalization of the converted results:

$$\pi = \frac{\gamma'}{\sum_i \gamma'_i}$$

Measures can be derived from π with the methods described in Section 3.4.

4 Examples

In this section two examples are given in order to illustrate the modeling power of the class of dtDSPNs.

Numerical results are given in Section 4.1 for a Geom/D/1/K - queueing system and in Section 4.2 for a D/D/1/K - queueing system with failures and repairs.

4.1 Geom/D/1/K - queueing system

The steady-state solution obtained for a Geom/D/1/K - queueing system with geometrically distributed arrivals, a single deterministic server, and a finite capacity is compared with a M/D/1/K - queueing system with exponentially distributed arrivals. Note that the geometric distribution approximates the exponential distribution in discrete time.

The solution for the M/D/1/K - queueing system is obtained with the solution method known for DSPNs [AC87] with the tool DSPNexpress [Lin92]. The Geom/D/1/K - queueing system and the M/D/1/K - queueing system were modeled with a dtDSPN and a DSPN, respectively. Figure 6 shows the dtDSPN model. In the according DSPN model the geometric transition is simply replaced by an exponential transition with the same mean firing time.

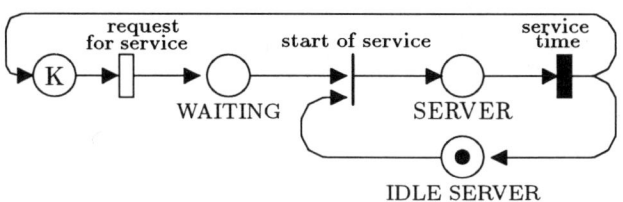

Fig. 6. The dtDSPN model of a Geom/D/1/K queueing system.

The time step of geometric firing time distribution is set to $\omega = 1$. The parameters are chosen according a capacity of 3 customers and of a service time of 4 time units. In a first experiment the mean number of customers is compared for a varying mean arrival time. The result is shown in Table 1. In a second experiment the distribution of customers is compared for a mean arrival time of 3.3 time units. The result is shown in Table 2. The results demonstrate a similar behaviour of both models.

mean arrival time [in time units]	mean number of customers			
	waiting		in service	
	dtDSPN	DSPN	dtDSPN	DSPN
10	0.09	0.11	0.4	0.4
5	0.44	0.47	0.74	0.72
3.3	0.91	0.90	0.92	0.89
2	1.46	1.41	1.0	1.0

Table 1:

number of customers	distribution of customers			
	waiting		in service	
	dtDSPN	DSPN	dtDSPN	DSPN
3	0.0	0.0	0.0	0.0
2	0.23	0.25	0.0	0.0
1	0.46	0.40	0.92	0.89
0	0.31	0.35	0.075	0.11

Table 2:

4.2 D/D/1/K - queueing system with failures and repairs

Consider a flexible manufacturing system with a facility for loading and processing of raw parts. The system has a buffer for K parts. Both the loading and the processing facilities work with a deterministic speed but are subject to failures and repairs. Figure 8 shows a dtDSPN model of the system. The model represents a D/D/1/K - queueing system with failures and repairs. Transition t_1 and t_2 correspond to the loading and processing of parts. Transitions t_3 (t_5) and t_4 (t_6) represent failure and repair of the loading (processing) facilities. Tokens in p_1 model free buffer places, tokens in p_2 model loaded parts. A token in p_3 (p_5) represents a failed loading (processing) facility and tokens in p_4 (p_6) represent a working loading (processing) facility, respectively. Arcs with a small circle at their destination are inhibitor arcs. The loading time is $Const(\tau_l)$, the processing time $Const(\tau_p)$. Both repair times are $Const(\tau_r)$. Failure times for the loading and processing facilities are $Geom(p_{f_{load}}, 1)$ and $Geom(p_{f_{proc}}, 1)$, respectively. We assume the following parameters: $K = 5, \tau_l = 1, \tau_p = 2, \tau_r = 3$, $p_{f_{load}} = 10^{-3}$. The underlying time step is therefore given by $\omega = 1$. Note that in most states two transitions with a deterministic firing time are enabled. In the following experiments the failure probability $p_{f_{proc}}$ is varied from 10^{-4} to 0.5. The model has 24 markings and 104 states. Figure 8 shows the throughput of the processing facility.

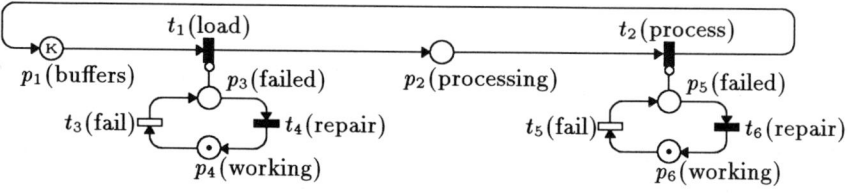

Fig. 7. D/D/1/K-queueing system with failures and repairs

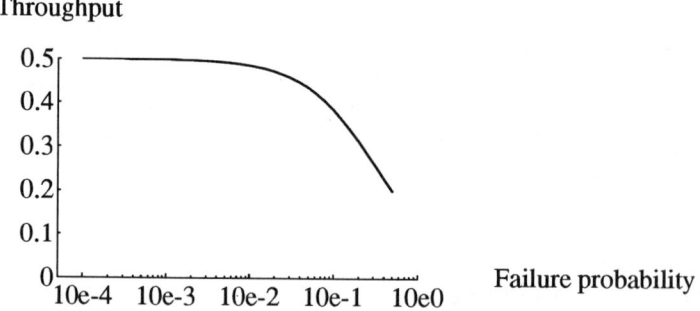

Fig. 8. Throughput of the processing facility

5 Conclusions

In this paper we defined the class of discrete time deterministic and stochastic Petri nets (dtDSPNs) for the modeling of systems comprising concurrent activities with either deterministic or geometrically distributed duration. No restrictions are imposed on the structure of the net. An algorithm was presented which computes the state space of the model and the one-step probabilities of the underlying DTMC. All conflicts can be resolved automatically by the algorithm. Transient and stationary measures can be computed by analyzing the DTMC. Although the state space may become large in case the greatest common divisor of the time steps of all firing time distributions is small, the approach is well suited to model technical systems where the most activities have a constant duration and failures can be modeled by random variables. State reduction techniques were proposed in addition. Numerical results were presented for the Geom/D/1/K and for a D/D/1/K queueing system with failures and repairs.

In future research we intend to include marking dependent and arbitrary discrete distributions of firing times into the formalism.

References

[ABC84] M. Ajmone Marsan, G. Balbo, G. Conte. "A Class of Generalized Stochastic Petri Nets for the Performance Analysis of Multiprocessor Systems". In *ACM Transactions on Computer Systems*, Vol.2, No.1, pp. 93-122, May 1984.

[AC87]   M. Ajmone Marsan, G. Chiola. "On Petri Nets with Deterministic and Exponentially Distributed Firing Times". In *G. Rozenberg (Ed.) Advances in Petri Nets 1987, Lecture Notes in Computer Science 266*, pp. 132-145, Springer 1987.

[Cia&al93] G. Ciardo, A. Blakemore, P. F. J. Chimento, J. K. Muppala, and K. S. Trivedi. "Automated generation and analysis of Markov reward models using Stochastic Reward Nets". In C. Meyer and R. J. Plemmons, editors, *Linear Algebra, Markov Chains, and Queueing Models*, volume 48 of *IMA Volumes in Mathematics and its Applications*, pages 145-191. Springer-Verlag, 1993.

[CGL93] G. Ciardo, R. German, C. Lindemann. "A characterization of the stochastic process underlying a stochastic Petri net". In *Proc. 5th Int. Conf. on Petri Nets and Performance Models*, Toulouse, France, October 1993.

[Çin75] E. Çinlar. *Introduction to Stochastic Processes*. Prentice-Hall, 1975.

[GL93]   R. German, C. Lindemann. "Analysis of Stochastic Petri Nets by the Method of Supplementary Variables". In *Proc. of the PERFORMANCE '93*, Rome, Italy, Sept.-Oct. 1993.

[GH85]   D. Gross, C. M. Harris. *Fundamentals of Queueing Theory*. 2nd Edition, John Wiley & Sons, 1985.

[Lin92] C. Lindemann. "DSPNexpress: A Software Package for the efficient Solution of Deterministic and Stochastic Petri Nets". In *Proc. 6th Int. Conf. on Modeling Techniques and Tools for Computer Performance Evaluations*, Edinburgh, Great Britain, September 1992.

[HV87]   M. A. Holliday, M. K. Vernon. "A Generalized Timed Petri Net Model for Performance Analysis". In *IEEE Transactions on Software Engineering*, Vol.SE-13, No.12, pp. 1297-1310, Dezember 1987.

[Mol85]   M. K. Molloy. "Discrete Time Stochastic Petri Nets". In *IEEE Transactions on Software Engineering*, Vol. 11, No. 4, pp. 417-423, April 1985.

[Mur89]   T. Murata. "Petri Nets: Properties, Analysis and Applications". In *Proc. of the IEEE*, Vol. 77, pp. 541-580, 1989.

[Zub91]   W. L. Zuberek. "Timed Petri Nets, Definitions, Properties, and Applications". In *Microelectronics and Reliability*, Vol. 31, pp. 627-644, 1991.

Bauhaus Linda: An Overview*

Nicholas Carriero, David Gelernter and Lenore Zuck

Department of Computer Science
Yale University
New Haven, Connecticut 06520

Abstract. Linda's success, based on a combination of simplicity and efficient implementation, encourages us to continue the quest for systems of "powerful simplicity." We introduce "Bauhaus Linda" (or Bauhaus for short), a Linda-derived coordination language that is in many ways simultaneously more powerful and simpler than Linda. Bauhaus unifies tuples and tuple spaces, leading to an especially clean treatment of multiple tuple spaces, and treats processes as atomic and explicitly representable. We present an informal semantics of Bauhaus and discuss an extended example that demonstrates its expressivity and simplicity.

1 Introduction

The Linda coordination language[CG92] represents one successful approach to parallel and coordinated programming. Linda's success seems to be due equally to two factors: the simplicity and power of the design and the efficiency of the implementation[2]. Despite its power, however, Linda lacks certain attributes that follow naturally from the tuple space model but do not figure in the current design. Two important examples are multiple first-class tuple spaces, and operators that apply in essentially the same way to data and process objects within a tuple space. In this paper we describe a new "Linda-like" coordination language that supplies these attributes in a framework that is (arguably) simpler than Linda itself.

The new system is called "Bauhaus Linda" (or just **Bauhaus**). Bauhaus generalizes Linda in the following ways:

1. *Linda's distinction between tuples and tuple spaces is eliminated.* Both are replaced by a single structure, the multiset (or mset). Instead of adding tuples to and reading or removing them from a tuple space, **Bauhaus**'s **out**, **rd** and **in** operations add multisets to and read or remove them from another multiset. Thus **Bauhaus** replaces Linda's ordered tuples with (unordered) multisets. Linda's tuple spaces are themselves already a type of multiset. By eliminating the distinction between tuples and tuple spaces, we make it possible in effect to

---

* Research of first two authors partially supported by Air Force Grant AFOSR-91-0098 and ONR-N00014-93-1-0573
[2] For example, a recent textbook describing an experiment in which Linda and the message-passing package PVM were compared on a seismic application describes Linda as "Elegantly simple to use. Seismic speedup comparable to PVM." [AG94]

treat a tuple space like a tuple: to create a tuple space, and to read or remove one in its entirety, via **out**, **rd** and **in**. We make it possible to treat a tuple like a tuple space: to add elements to, and read or remove individual elements from, a tuple. By specifying that tuples and tuple spaces are both superceded by (in particular) multisets, we suggest how a tuple or tuple space might be treated once it has been bound as a consequence of a **rd** or **in** operation to a local variable. Operations that are appropriate over multisets—for example union, intersection, subtraction, iteration over all elements—will be appropriate to locally bound tuples and tuple sets as well.

2. *Linda's distinction between tuples and anti-tuples is eliminated.* In Linda, **out** specifies a tuple, **in** and **rd** an anti-tuple (or "template"). Formal parameters may in principle appear within either, but in practice formal parameters are used in anti-tuples only, and tuples and anti-tuples have come to be formally distinct. In Bauhaus on the other hand, **out**, **rd** and **in** each take a multiset argument. The **out** operation adds the specified multiset to some target multiset. The **rd** and **in** operations return any multiset that contains the specified multiset. Thus Linda's type- and position-sensitive associative matching rule is replaced by simple set-inclusion. (A tuple-matching rule based on set inclusion is obviously appropriate to a coordination language based entirely on multisets.)

3. *Linda's distinction between active data objects (processes) and passive data objects is eliminated.* Processes and passive objects remain distinct species in Bauhaus—a process acts and a passive object does not—but the coordination language *itself* makes no distinction between them. Thus in Linda, processes can only be added to a tuple space via **eval**, passive objects only via **out**; in Bauhaus, **out** serves for both. In Linda, only passive objects can be read or removed from a tuple space; in Bauhaus, both processes and passive data may be read or removed. (*Reading* a process results in the reader's acquiring a suspended copy of the process image. *Removing* a process results in the remover's acquiring the active process itself. Thus neither the **rd** nor the **in** operation in itself alters the number of active processes in the system.)

Bauhaus is thus substantially simpler than Linda, in the sense that it makes no distinction between tuples and tuple spaces, tuples and anti-tuples, processes and passive data objects; and Bauhaus makes do with three primitive operations where Linda requires four. The simplifications in Bauhaus make for substantial added power:

1. *Because Bauhaus makes no distinction between tuples and tuple spaces, it naturally accomodates a hierarchy of multiple tuple spaces.* ("Classical" Linda supports only one global tuple space; other Lindas support multiple tuple spaces, although generally within a single flat meta-space and not in a hierarchy—e.g. Paradise [Par94], TS[Jag92], PROSET-Linda [Has91].) In Bauhaus, **out** adds a new multiset N to some target multiset; but N can just as well be treated as a *tuple space* as a tuple. In other words, **out** can be used to add either a new tuple *or* a new tuple space to some target space. Note that the existence of multiple tuple spaces will, on the other hand, require Bauhaus to acquire a piece of apparatus that Classical Linda does not need: some way to designate the target

multiset of an **in**, **rd** or **out** operation (in other words, to specify which tuple space is currently of interest).

2. *Because tuples and tuple spaces are replaced by multisets, and formal parameters are eliminated in favor of* **in** *and* **rd** *operations that yield multisets as values,* **Bauhaus**'s *multiple tuple spaces are first class.* A first-class value can be yielded as the result of an expression and passed as an argument to a function. In Bauhaus, **in** and **rd** yield multisets (i.e. tuple spaces), which are bound to local variables of type multiset, and which may in turn be operated upon locally (the reading or removing process might iterate over all members of a bound tuple space, print it, pass it some function) or dropped into some other tuple space, thus in effect be "transplanted" into a foreign context. We assume set manipulation operations, such as those found in languages like SETL [Sch86], would be made available to work with the multisets bound to local variables.

3. *Because the distinction between processes and passive objects is eliminated, and given the hierarchy of multiple tuple spaces discussed above,* **Bauhaus** *can model an unusually wide spectrum of coordinated systems.* A *Turingware* system is one in which people (or their software proxies) and "ordinary" processes (corresponding to or modelling no particular person) are embedded in the same coordination framework [Gel91]. Turingware is a useful idea because it allows us to see phenomena hitherto evidently unrelated (for example, a groupware editor and a human-computer ensemble) as instances of the same structure, and because it allows us to interpolate new structures (an instance of the software architecture called "Trellis," for example, in which some decision procedures are implemented by processes and some by people) where previously there were conceptual holes. Bauhaus is well-suited to the needs of Turingware where Linda is not: in Bauhaus, for example, a person-proxy can travel from tuple space to tuple space, interacting with the local environment at each stage of the game. In Bauhaus, two people-proxies wanting to communicate can create a new tuple space exactly for that purpose, and both hop in. Bauhaus can serve, also, as a model for an arbitrarily-large chunk of the world at large (which is, after all, just a big coordination framework with lots of embedded computations). Bauhaus can model the Internet, a unix file system or a conventional (non-electronic) library. Using Bauhaus in this way has the same general advantages of pursuing Turingware: we see how seemingly-disjoint phenomean are related, and how we might invent new coordination frameworks and ensembles.

In the sections following, we present an informal but reasonably precise Bauhaus semantics; discuss an extended example; and conclude with a brief comment on implementation.

2 An Informal Semantics for Bauhaus

Let A be a universal set of *atomic elements*. An *mset* is either a single A-element or a (possibly empty) multiset of msets. A state of a **Bauhaus** system is described by the *coordination structure*, the "outermost" mset and it's contents, and the states of the processes in the coordination structure. The coordination structure has an obvious tree representation; we will use msets and trees interchangeably in the following. An interior node of the tree represents an mset which is an element of the mset described by its parent node, and which contains as elements the msets described by its children in the tree. Atomic elements (and null msets) are the leaves of the tree. Atomic elements include passive data (values like "5" and the string "cinco") and active data (processes undergoing evaluation). If no element in an mset is a running process we say the mset is *dead*.

The processes in a **Bauhaus** structure manipulate the structure by executing **Bauhaus** commands. If necessary, we introduce the type `mset` and mset constructors into the hosting language.

Each **Bauhaus** command has two mset arguments, *where* and *what*. The where-argument designates a non-atomic mset with respect to which the operation takes place. The what-argument gives the operand. In the case of `in` and `rd`, the what-argument is the search template for the sought after mset. In the case of `out`, the what-argument is the mset to be inserted.

Where-arguments define msets that are to be embedded (in the graph theoretic sense) in the **Bauhaus** coordination structure. A where-argument contains two distinquished atomic elements, $\bigcirc$ and $\rightarrow$. The tree corresponding to the where-argument is embedded in the tree of the coordination structure, with the $\bigcirc$ node anchored at the leaf corresponding to the executing process. Once embedded, the $\rightarrow$ of the where tree will then label a non-atomic mset in the coordination structure. (If it fails to do so, the operation suspends until the coordination structure is modified in such a way that a non-atomic node is labelled.) The operation will take place with respect to the labelled mset.

For example, used as a where-argument,

$$\{\bigcirc, \{\rightarrow \{a, \{\}, \{\}\}\}\}$$

locates "a child of my sibling that has at least three children, one of whom is an a-node and both the others have children."

The what-argument may not contain the distinguished elements. The node selected by an `in` and `rd` command is a child of the node designated by the where-argument that can embed the what-argument. For example, used as a what-argument in a `in` or `rd` command,

$$\{\{a, c\}, \{b, c\}\}$$

should select a node that has one child that has a- and c-children, and another child that has b- and c-children. For the `out` command, the what-argument is inserted as a new child of the node designated by the where-argument.

An example of a complete command is

$$\text{in } \{\bigcirc, \rightarrow \{\{a\}, \{b\}\}\}, \ \{\{c\}\}.$$

This command locates a sibling of the issuing process that has a child that has an a-child and a child that has a b-child, and removes (and returns to the issuing process) one of the sibling's children that has a c-child.

Notes

- The msets described by where-arguments (and what-arguments of **in** and **rd**) are non-deterministic; there may be several nodes that satisfy them. Obviously, it is also possible that they cannot be satisfied. If a node satisfying the where-argument of an **in** or **rd** fails to satisfy the what-argument, another node satisfying the where-argument may be tried.
- It is, in practice, impossible to obtain a "still picture" of the coordination state of the system. We do, however, require that if some node that satisfies the argument(s) is found, then there is a snapshot of the system that leads to this node and does not violate causality. In other words, we require that the is a consistent global state of the system (see [CL85, Mat89, Mor85] for more on "consistent global states") which satisfied the argument(s) and points to the returned node.
- Some time may elapse between the time a good node is found and the time the issuing process receives that node. Hence, it is conceivable that a returned node is no longer a "good" node (i.e., it is not a child of a node led to by the where-argument, or it does not satisfy the what-argument) when it is received by the issuing process. A process may want to know whether the returned node is good or not. We can make available a mechanism that detects that. This can be done, for example, by timestamping nodes that are accessed when searching for a good embedding of the argument(s).

The out command

An **out** command locates a node specified by its where-argument, and inserts the mset described by the what-argument as a new child of the located node. For example, if the **Bauhaus** coordination state is

$$\{P, \{a, b\}\}$$

and the process at the node marked by P executes

$$\textbf{out} \rightarrow \{\bigcirc\}, c,$$

then the **Bauhaus** coordination state becomes

$$\{P, \{a, b\}, c\}.$$

If the process at P then executes

$$\textbf{out} \{\bigcirc, \rightarrow \{\}\}, c,$$

the **Bauhaus** coordination state becomes

$$\{P, \{a, b, c\}, c\}.$$

Notes

- A Bauhaus **out** command may block when there is no embedding of the the where-argument.
- The what-argument can be, of course, an mset-typed variable. In this case, the value of the variable will be inserted into the structure.
- Once an mset is inserted into the Bauhaus structure, the process that issued the **out** command loses the mset. In particular, it no longer "owns" the processes in the mset.

The in command

An **in** command locates an mset specified by its where-argument. It then finds one of the elements of this mset that satisfies the what-argument, removes that element from the Bauhaus coordination state and returns it to the issuing process.

For example, if the Bauhaus coordination state is

$$\{P, \{a, b, c\}, c\}$$

and P executes

$$\mathtt{m = in} \ \rightarrow \{\bigcirc\}, \ \mathtt{c},$$

(where m is of type **mset**), then after the command returns, m is c and the Bauhaus coordination state is

$$\{P, \{a, b, c\}\}.$$

Notes

- The **in** command may block for two reasons: there is no node labelled by the where-argument, or there is no child that matches the template described by the what-argument. We require that if the command does not block, its result is a node that satisfies both embeddings in the sense described above. We also require that the command does not block if there is an embedding that satisfies the arguments that is consistently in the Bauhaus coordination state.
- The mset that is returned to the issuing process is removed from the Bauhaus coordination state. In particular, from some point on, the "live" processes in it do not have access to the complete mset that describes the Bauhaus coordination state. Similarly, processes outside the returned mset cannot access the msets nested within the **in**ed mset from this point on.

The rd command

An **rd** command locates an mset specified by its where-argument. It then finds one of the elements of this mset that satisfies the what-argument, and produces a "dead copy" of the mset to be returned to the issuing process. By a "dead copy" we mean a copy where process nodes are replaced by a memory snapshot of the process nodes.

For example, if the **Bauhaus** coordination state is

$$\{P, \{a, b, c, Q\}, c\}$$

and P executes

$$\mathtt{m = rd} \;\rightarrow \{\bigcirc\}, \;\{\mathtt{c}\}$$

(where m is of type mset), then after the command returns, m is $\{a, b, Q', c\}$ where Q' is a snaphot of Q, and the **Bauhaus** coordination state remains intact.

Notes

- The rd command may block exactly when the in command may block. Similarly to the in command, we require that the command does not block if there is an embedding that satisfies the arguments that is consistently in the **Bauhaus** coordination state.
- The mset that is selected by the issuing process is not removed from the **Bauhaus** coordination state.
- The node which is copied may contain processes that may change by the time the copy of the node is returned.

3 Example

We use syntax in this example that is informal rather than precise. Our intent is to make clear the capabilities of the system.

3.1 An Electronic Library

Suppose our goal is to manage a library of online documents. The documents might by ASCII text or scanned-in images; we assume they are the sort of documents that people find in conventional libraries—books, journals and so on. Clearly a conventional library, not a database, will be our model for managing this sort of document collection.

A book would take the form

    {{BOOK}, {LOG}}.

(capitalized names will designate a collection of elements.) The first element of BOOK is a title,

    {"U.S.A.", Index, {KEYWORDS}, ...};

you make a copy of this book for your own use (you "check it out") by executing a rd operation—for example,

    MyCopy = rd {{"U.S.A."}}.

This rd operation—which assumes that you are inside "Books", a multiset which looks like

{"Books", { ... }, ..., {{"U.S.A.", ...}, ...}, ... },

and omits the where argument—means "read some multiset containing a superset of "U.S.A."." (Note that this operation would not have been possible in Linda, because it requires reading an entire mset—in effect a whole tuple space; this mset might furthermore contain active processes (as we discuss below), which can't be rded or ined in Linda.

Elsewhere (in some other mset), tables are stored in which authors, keywords and so forth are mapped to indices. The user acquires a copy of the book with index n by executing

MyCopy = rd{{n}}.

The collection of elements designated LOG holds a stream of comments about the book: corrections or comments by the author (if he happens to be alive and online), questions or comments by readers, scholarly notes by researchers or librarians bearing on the relationship of this book to others; possibly reviews or other pieces that discuss the book. The stream of comments is represented as it would be in Linda, as a collection of numbered msets:

{1, comment1}, {2, comment2}, ... {"next", pointer}

To append to the stream, a user removes the "next" mset using in, increments the value of pointer and replaces the set using out. He then outs a new mset holding the text of his comment and the pre-increment integer value of pointer.

Suppose a user is interested in books A, B and C. He might choose to toss a "keep me informed" daemon into the LOG mset of each book. Each daemon sits astride one comment stream and alerts the user whenever new comments are appended; the user is kept informed of new developments around the book. The daemons execute

```
while (1)
{  NewComment = rd{next++};
   print or store NewComment  }
```

(Each comment stream might be seen as an instance of the *Lifestreams* discussed in [FCG94].)

Now, suppose a user has built an mset containing books of interest to him:

{{BOOK1}, {BOOK2},...}

He will probably want to sort it before displaying it as, for example, a list of titles on his screen. To carry out the sorting, he would toss a process into this mset. The process would execute a function designed to do the desired kind of sorting: by closeness of keyword match, publication date, author, etc. These functions acquire a context (figure out, in other words, which mset they are supposed to be sorting) simply by dint of being tossed into the right mset.

Suppose a user wants to be informed whenever the library acquires a new book about trampolines. Index numbers will be assigned in order, and so a process can examine each new acquisition by finding out (by referring to the appropriate datum in the "catalog" mset) the index of the most recently-acquired book (say n), then awaiting for $n + 1$ to arrive, then $n + 2$ and so on:

```
while (1)
{   NewBook = rd{{n++}};
    check NewBook's keywords... }
```

Suppose Schwartz is coming to campus to give a lecture, and the sponsors want to inform interested parties. They can create a "PR Daemon" that attaches a notice to the comment streams associated with each book by Schwartz.

This observation is minor in itself, but leads to a more important one: in our Bauhaus-built electronic library, books are not merely books; they are information exchanges or marketplaces. Books become natural foci for electronic discussion of the topics with which they deal. Most books represent, of course, finely-tuned mixtures of subjects and attitudes—far more fine-tuned, for example, than the designation "comp.parallel" (the name of an internet bulletin board). As such they represent, potentially, a sensitive and subtle way to place people with like interests in communication. In this *Weltanschauung*, books become more and not less important as the world goes electronic.

Suppose a user has compiled a large mset of interesting books and wants to perform a complex search over the texts—say, to produce a complete cross reference index. This process could take some time on his desk-top machine. It would be desirable if he could execute it in parallel instead.

Our electronic library might provide exactly such a parallel-execution capacity in the form of an mset called "TurboSearch." TurboSearch would include a number of processes, each associated with its own processor. (The processors might be part of a multiprocessor or might be nodes in a LAN.) The user could acquire the TurboSearch resource simply by ining it, which gives him exclusive access to the processes and processors in TurboSearch for as long as he needs them. Assume that TurboSearch should be handed around to users in the order in which they post their requests. A user first acquires and increments a token. If the token is stored in a set of the form

```
{"TurboSearch", "token", n},
```

the user executes an **in** `{"TurboSearch", "token"}` operation, examines n and then executes **out** `{"TurboSearch", "token", n+1}`. He then waits for the value of the **NextUser** to progress to n:

```
in {"TurboSearch", "NextUser", n}.
```

When he continues subsequent to this **in** operation, he grabs the resource—

```
MP = in {TurboSearch},
```

manipulates **MP** locally (inserting his request and retrieving the answer), and then frees the resource and gives the next waiting user (if there is one) a chance:

```
out MP;
out {"TurboSearch", "NextUser", n+1}.
```

4 Conclusions

By eliminating the distinction between tuples and tuple spaces, tuples and antituples, and active and passive data objects, **Bauhaus** becomes in some ways a significant simplification of Linda. It is arguably more expressive as well. Because simplicity and expressive power have always been our main goals, **Bauhaus** may be a natural successor to Linda in some areas.

On the other hand, Linda in certain ways is simpler than **Bauhaus**; and Linda's efficient, highly-optimized family of implementations has been crucial to her success. We have learned by experience that it is foolish to dismiss any language as incapable of efficient implementation before serious investigation has taken place; it is possible that **Bauhaus** can be implemented as efficiently (or for all we know *more* efficiently) than Linda. But it is clear that efficient implementation of **Bauhaus** poses a number of hard problems. In the meantime, **Bauhaus** may prove useful in areas (for example Turingware and some distributed applications) where efficiency is less important than robustness and expressivity.

References

[AG94] G. Almasi and A. Gottlieb. *Highly Parallel Computing, Second Edition.* Benjamin/Cummings, Redwood City, CA, 1994.

[CG92] N. Carriero and D. Gelernter. Coordination languages and their significance. *Commun. ACM*, 35(2):97–107, Feb. 1992.

[CL85] K. M. Chandy and L. Lamport. Distributed snapshots: Determining global states of distributed systems. *ACM Trans. Comput. Syst.*, 3(1):63–75, 1985.

[FCG94] E. Freeman, N. Carriero, and D. Gelernter. Getting rid of windows, files, desktops, email, names and compatibility: an introduction to Lifestreams. Technical report, Yale University Department of Computer Science, Aug. 1994.

[Gel91] D. H. Gelernter. *Mirror Worlds.* Oxford, New York, 1991.

[Has91] W. Hasselbring. On integrating generative communication into the prototyping language PROSET. Informatik-Bericht 05-91, Universität Essen, Dec. 1991.

[Jag92] S. Jagannathan. TS/Scheme: Distributed Data Structures in Lisp. In *2nd Workshop on Parallel Lisp: Languages, Applications and Systems*, Oct. 1992. Also published as: NEC Research Institute Tech Report: 93-042-3-0050-1.

[Mat89] F. Mattern. Virtual time and global states of distributed systems. In M. Cosnard, editor, *Proceedings of the International Workshop on Parallel and Distributed Algorithms*, pages 215–226. North-Holland, 1989.

[Mor85] C. Morgan. Global and logical time in distributed algorithms. *Information Processing Letters*, 20:189–194, 1985.

[Par94] Scientific Computing Associates, New Haven, CT. *Paradise User's Guide*, 1994.

[Sch86] J. T. Schwartz, et.al. *Programming with sets: an introduction to SETL.* Texts and monographs in computer science. Springer-Verlag, New York, 1986.

Naming and typing in languages for coordination in open distributed systems

Robert Tolksdorf

Technische Universität Berlin
Fachbereich 13, Informatik
Funktionales und Logisches Programmieren
Sekr. FR 6–10
Franklinstr. 28/29, D-10587 Berlin
e-mail: tolk@cs.tu-berlin.de

Abstract. Open distributed systems have to take into account a number of heterogeneities in the systems components and possibly high dynamics in the systems structure by unrestrictedly joining and leaving agents. Solutions to the coordination problem in open systems should be based on the notion of services, which are provided and used by agents.
For the identification of services, typed interfaces provide an appropriate mechanism. However, a crucial issue in designing such an identification schema is the usage of names. We investigate in alternatives on dealing with names that take into account the requirements of open systems.
We demonstrate that is may well be of benefit to introduce other mechanisms than syntactical matching on names such as their replacement by structural information or their complete exclusion. We use this insight to define a type system with subtyping that pays special attention to a flexible treatment of names.
This type system is then applied to the coordination language for open systems LAURA. Here, services are identified by the definition of an interface consisting of operation signatures. We give semantics to these interfaces with the type system by discarding all names in arguments and results and by applying syntactical name-matching on operation-names only.

1 Introduction

Open distributed systems are information systems that pose special requirements for the coordination of agents forming them. These systems have to deal with a number of heterogeneities such as the usage of different hardware-architectures, of different network-topologies and -protocols, and of a variety of different programming languages used to implement agents. They have a potential world-wide scale and operate continuously without a defined beginning or end. Besides heterogeneity, a main characteristic is the dynamic behavior of such systems, as agents can join and leave them without restrictions at any time. Figure 1 illustrates the environment for an open system with various different machine- and network-architectures.

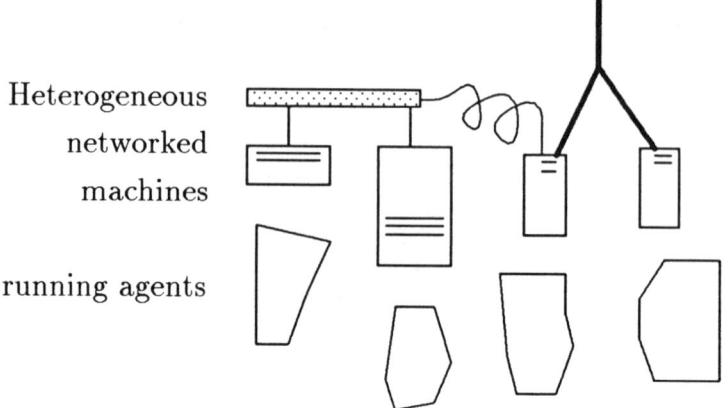

Fig. 1.: Components of an open system, missing is the coordination "glue".

A common understanding is that open distributed systems are based on services that are requested and offered by agents. That is, an agent is able to perform actions as a service-provider that are of benefit for another agent, the service-user.

As an example, let us look at a hypothetical open system. Say, travel agencies and carriers want to set up a system that enables customers to make travel bookings from their personal computers at home. Payments made by credit cards have to be authorized by banks. Such a system is appealing as nearly all components involved already exist, as bookings probably can be made faster and easier and as some savings and gains are expected by the involved companies[1].

Components involved can include personal computers at the home of potential customers, a mainframe running a transaction system at a bank and workstations running a distributed reservation system. Connections can be established by low-band ISDN, by dial-up or permanent telecom-lines or inhouse LANs.

Offers and reservations are transmitted to the carriers using a reservation system and thus connected to the proprietary reservation systems of the carriers. Finally, authorization of credit cards is possible automatically from automata connected to some bank or via an interactive query. Charging credit cards results changes in the customers accounts which take place in some banking system.

All software-components can be implemented in very different programming languages. There could be some SQL-driven database of travel-offers and some accounting program written in Cobol. A user interface for the customer could be written in C.

Finally, dynamics arise from customers joining the system at some time to make a booking. Agencies can be established or go out of business, as can carriers.

---

[1] Which, however will not lower prices for traveling!

The problem that arises is how to coordinate such a service. There are actions that constitute the service-provision – such as the reservation of a flight – and actions for communication and synchronization that are necessary for service-coordination in a parallel and distributed world of agents. Some "glue" has to be provided to put components together enabling them cooperate. We defined a language, LAURA ([Tolksdorf, 93b, Tolksdorf, 94]) for that purpose, which focuses on the coordination of services in open distributed systems.

LAURA sets up a *service-space* which is used for the coordination of services. The service-space contains *forms* that are added and retrieved by agents using LAURA's three operations. There are three kinds of forms: service-forms, serve-forms and result-forms.

A serve-form is put into the service-space by an agent performing a `serve`-operation and consists of a description of the offered service. It is returned to the offering agent with an operation from the service selected and a list of arguments for that operation filled out.

The service-provider then executes the operation and puts a result-form with `result` into the service-space in which some result fields are filled with the results of the operation.

A service-form is used by an agent executing a `service`-operation. Here, the requested service is described, an operation selected and a list of arguments provided. It is returned to the agent with result-fields filled out. In figure 2 such a service space in use by agents is depicted.

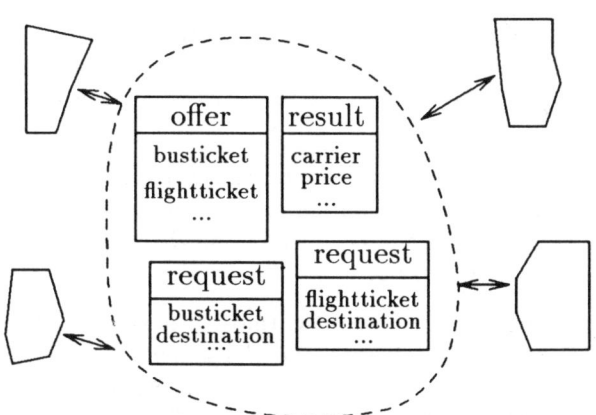

Fig. 2.: A service-space in LAURA for the traveling example

In this paper, we do not detail out LAURA's operations but pay special attention to one crucial aspect of the language. All kinds of forms contain a service-identification, describing what service is requested or offered. In this paper, we focus on how such a service-identification can be realized in the light of the requirements of open distributed systems.

This paper is organized as follows. In section 2 we describe how services are identified in LAURA by service-types. Then, we investigate in the issue of naming in open distributed systems in section 3 and define in section 4 a type-system that is special in that it offers a flexible choice on how to deal with names. Finally, we give a definition of LAURA's type definition language in section 5 which takes advantage of our type system.

2 Identification of services

The "glue" LAURA uses to coordinate services provided and used by agents in an open distributed system is the exchange of forms via the service-space. As put, forms identify the service requested or offered and necessary information. The question is how to identify a service.

In LAURA, a service is described as an interface consisting of a set of operation signatures. The signatures describe the types of the operation in terms of their names and their argument- and result-types. It is therefore a record of function-types. A form contains a description of this interface-type as the entry for service-identification. Putting a service-request form into the service-space causes the search for a service-offer form so that the interface-type of the offer is in a matching relation to that of the request. We do not introduce names for service-types, which is different to what is done in other approaches, such as direct naming and managed types.

In the first, a set of names is defined for services. An example are UNIX system-services that are "named" numerical by some port-number under which services are provided. There is a global convention that, for example, the numeric name 25 identifies the mail-service.

Such an identification scheme is static, that is the names have to be universal known in advance. ActorSpaces ([Agha and Callsen, 93]) also use a direct naming system, but allow regular expressions to be used for the identification of services. Let some services be named "mail", "mail-fast" and "simple-mail", then the expression "*mail*" identifies one out of those three. However, the name-portion "mail" still has to be known in advance. Such a scheme cannot be well-suited for open systems as they are dynamic in nature and as one cannot assume identical naming schemes at a world-wide scale. Our approach is different in that we do not use names for services, but identify them by their interface type solely.

A more dynamic scheme is defined for the ISO standard on open distributed processing ODP ([ISO/IEC JTC1/SC21/WG7, 93, ISO/IEC JTC1/SC21/WG7,

94a, ISO/IEC JTC1/SC21/WG7, 94b, ISO/IEC JTC1/SC21/WG7, 91][2]). Here, a repository of type definitions is defined which is used to store interface types of services and relations amongst them. A subtype-relation amongst interface types can be explicitly declared or derived from subtyping rules.

Offering or requesting a service is done using a trading function ([ISO/IEC JTC1/SC21, 93]) that stores offers and their types. It uses the type repository to determine relations amongst offered and requested types and provides a requestor with an identifier of an object that offers an appropriate service. Our approach differs from the ODP schema in that we do not introduce a repository of types.

In LAURA, services are identified by giving a service-type description only. LAURA's type-system as defined below allows us to infer from two descriptions, if an offered service matches a requested one. It does so, if it is a subtype of the request according to the type-system.

Subtyping in this system is defined so that a type A is a subtype of B if all values of type A can be substituted when a value of type B is requested. The "values" we type are services. The subtyping enables us to use a service of type A if a service of type B is requested.

For the traveling example, the typing makes it possible to have an agency that offers bus-, train- and flight-tickets perform the purchase of a train-ticket when an agency is requested that offers bus- and train-tickets. It also rules out agencies offering bus- and flight-tickets to be selected for the purchase.

LAURA's type system is given as a set of inference rules for type-equivalence and subtyping. It includes simple types and a set of constructed types such as records and unions. LAURA's service-types are given semantics by defining their interpretation in the type language.

However, before formalizing the type-language, we have to discuss the issue of naming in open distributed systems in more detail, it will impose special problems, when speaking about named operations that form a service.

3 Naming in open distributed systems

Above we stated informally that a service interface type consists of a number of operation signatures formed by an operation name and lists of types for arguments and results. A relation on operation signatures will take the operation names into account. An interface signature then can be understood as a record consisting of tagged fields, where the tag corresponds to the operation name and the value to the type of the function.

---

[2] ODP is a standard which is in the ballot phase. The references we give in this paper refer to those documents that were available at the time of writing. There may be changes in the further stages of standardization that could lead to deviations from our statements. ODP parts 2 and 3 are expected to become international standards in 1995, parts 1 and 4 are expected to reach this status in 1996 ([Raymond, 94]). ODP will become the ISO/IEC standard 10746 and ITU-T recommendations X.901 to X.904.

Common equivalence relations for tagged records require all tags and their associated values to be equal. A subtyping-relation as in [Cardelli, 88, Amadio and Cardelli, 91], requires that all fields of the supertype have to have corresponding fields in the subtype with identical tags and values being in a subtype relation to those of the supertype. Examing such a rule closer shows that typechecking involves some form of name checking, too. In fact, for the given rule this name-checking means testing for syntactic equivalence.

Let us assume that flight-carriers register some information about a traveler at check-in. At Heathrow Airport, London, this information is kept in a record $A = \langle$baggagetag: string, withchild: boolean, smoker: boolean$\rangle$. For a transatlantic flight this information is passed to the destination airport, which could be JFK, New York. Information about passengers there is stored in records of the type $B = \langle$luggagetag: string, withchild: boolean$\rangle$.

We have to test, if A conforms to B when the information is to be transferred. A short glance at the record makes it clear that it should conform as we can forget the entry about smoking and as we know that "baggagetag" and "luggagetag" mean the same thing. Applying formal subtyping rules, however, does not identify A as being a subtype of B as the names are syntactical distinct. There is a discrepancy between what we informally state and what is yielded by the formal test.

We can represent a naming system by relations between names and semantic objects, where objects are represented by names. We call this relation $\mathcal{R}$ and write $n\mathcal{R}o$ when the name n represents the object o.

Names can be understood as an encoding of the intended meaning in identifying objects and the objects themselves as the extension of this identification. When we consider a naming system in which all semantic objects are identified by one single name and a name always identifies exactly one semantic object, we can justify a formal rule that requires a syntactical equivalence on names.

The example above illustrates that this assumption can be too restrictive as it disallows "baggagetag" and "luggagetag" to identify the same semantic object. For an open system in which global and consistent knowledge is absent, the same names can well be used for different semantic objects and different names can refer to the same object.

Let $\mathcal{M}$ be a relation on names that is used to determine if two names refer to the same semantic object. Four possibilities in relation to $\mathcal{R}$ arise for objects o_1 and o_2 and names n and m as follows:

- Accidental mismatch $\quad m\mathcal{R}o_1 \wedge n\mathcal{R}o_1 \wedge \neg(m\mathcal{M}n)$
- Intended match $\quad m\mathcal{R}o_1 \wedge n\mathcal{R}o_1 \wedge m\mathcal{M}n$
- Intended mismatch $\quad m\mathcal{R}o_1 \wedge n\mathcal{R}o_2 \wedge \neg(m\mathcal{M}n)$
- Accidental match $\quad m\mathcal{R}o_1 \wedge n\mathcal{R}o_2 \wedge m\mathcal{M}n$

We can define two properties: $\mathcal{M}$ is said to be *sound* if $m\mathcal{M}n \Rightarrow m\mathcal{R}o \wedge n\mathcal{R}o$ and to be *complete* if $m\mathcal{R}o \wedge n\mathcal{R}o \Rightarrow m\mathcal{M}n$. Soundness states that no accidental match will occur and completeness that no accidental mismatch occurs.

$\mathcal{R}$ is an abstract relation, today we have no means to derive $\mathcal{M}$ from it by some algorithm from a given set of names and objects. However, we can reason about the appropriateness of relations on names. We identify three outforms of $\mathcal{M}$ that seem reasonable:

1. $n\mathcal{M}m \Leftrightarrow n = m$. This is the syntactic relation on names which we referred to above.
2. $n\mathcal{M}m \Leftrightarrow s(n) = s(m)$. Here, s is a function on names that can provide some structural information. For a record, s could be defined as the position of a names field.
3. $n\mathcal{M}m$. Here, names are ignored by defining all names as pairwise matching.

Note that all three forms are not sound, as accidental matches can occur. The last relation is complete, as no accidental mismatches can occur. The difference between form one and three represents the semantics of names, the difference between two and three the semantics of ordering.

For LAURA, we chose the first form for interface-types where the names of operations offered have to be syntactical equivalent. We will use the last outform for any names that occur in the types used as arguments or results for operations. This decision is a compromise between completeness – wanting to accept the flight-information example – and soundness – avoiding to increase the quantitative amount of accidental matches.

Before formalizing this choice in the semantics of LAURA's, we define in the following the type system we use for this definition and which is special in that it contains rules that take into account the three relations on name matching we found reasonable.

4 A type system with subtyping

The type-system we use for LAURA is defined by a set of inference rules for type equivalence- and subtyping-relations. It consists of constants, three record-types with different matching for names, products, three union-types with different matchings for names and function types. Type terms are taken from a language which is generated by the following grammar:

$$\alpha ::= t \mid \langle a_1: \alpha_1, \ldots, a_n: \alpha_n \rangle \mid \langle a_1: \alpha_1, \ldots, a_n: \alpha_n \rangle_O \mid \langle a_1: \alpha_1, \ldots, a_n: \alpha_n \rangle_A \mid$$
$$[a_1: \alpha_1, \ldots, a_n: \alpha_n] \mid [a_1: \alpha_1, \ldots, a_n: \alpha_n]_O \mid [a_1: \alpha_1, \ldots, a_n: \alpha_n]_A \mid$$
$$\alpha_1 \times \ldots \times \alpha_n \mid \alpha \to \beta.$$

The different outforms of records and unions reflect the different forms of name-matching as described in the previous section. $\langle a_1: \alpha_1, \ldots, a_n: \alpha_n \rangle$ is an *exact record*, in which the names are used for syntactic matching, $\langle a_1: \alpha_1, \ldots, a_n: \alpha_n \rangle_O$ an *ordered record*, where the names are unimportant and only the structural information on the order of fields is used. When working with an *anonymous record* – $\langle a_1: \alpha_1, \ldots, a_n: \alpha_n \rangle_A$ – both the syntactic and structural properties of the names are neglected. The rules below will make it clearer what exactly this means.

4.1 Rules for type-equivalence

Equivalence of types is defined by the inference rules in figure 3. They establish an equivalence relation $=$, which is reflexive (EREFL), symmetric (ESYM) and transitive (ETRAN).

$$\frac{}{\alpha = \alpha} \text{EREFL} \qquad \frac{\alpha = \beta}{\beta = \alpha} \text{ESYM} \qquad \frac{\alpha = \beta \quad \beta = \gamma}{\alpha = \gamma} \text{ETRAN}$$

$$\frac{\forall i \in \{1, \ldots, n\} : \alpha_i = \beta_i}{\langle a_1 : \alpha_1, \ldots, a_n : \alpha_n \rangle = \langle a_1 : \beta_1, \ldots, a_n : \beta_n \rangle} \text{ERECEX}$$

$$\frac{\langle 1 : \alpha_1, \ldots, n : \alpha_n \rangle = \langle 1 : \beta_1, \ldots, n : \beta_n \rangle}{\langle a_1 : \alpha_1, \ldots, a_n : \alpha_n \rangle_O = \langle b_1 : \beta_1, \ldots, b_n : \beta_n \rangle_O} \text{ERECORD}$$

$$\frac{\langle a_x : \alpha_x, \ldots, a_y : \alpha_y \rangle_O = \langle b_1 : \beta_1, \ldots, b_n : \beta_n \rangle_O \quad \langle a_x : \alpha_x, \ldots, a_y : \alpha_y \rangle \in \binom{\langle a_1 : \alpha_1, \ldots, a_m : \alpha_m \rangle_O}{\langle a_{i_1} : \alpha_{i_1}, \ldots, a_{i_m} : \alpha_{i_m} \rangle_O}}{\langle a_1 : \alpha_1, \ldots, a_n : \alpha_n \rangle_A = \langle b_1 : \beta_1, \ldots, b_n : \beta_n \rangle_A} \text{ERECANON}$$

$$\frac{\langle 1 : \alpha_1, \ldots, n : \alpha_n \rangle_A = \langle 1 : \beta_1, \ldots, n : \beta_n \rangle_A}{\alpha_1 \times \ldots \times \alpha_n = \beta_1 \times \ldots \times \beta_n} \text{EPROD}$$

$$\frac{\forall i \in \{1, \ldots, n\} : \alpha_i = \beta_i}{[a_1 : \alpha_1, \ldots, a_n : \alpha_n] = [a_1 : \beta_1, \ldots, a_n : \beta_n]} \text{EUNIEX}$$

$$\frac{[1 : \alpha_1, \ldots, n : \alpha_n] = [1 : \beta_1, \ldots, n : \beta_n]}{[a_1 : \alpha_1, \ldots, a_n : \alpha_n]_O = [b_1 : \beta_1, \ldots, b_n : \beta_n]_O} \text{EUNIORD}$$

$$\frac{[a_x : \alpha_x, \ldots, a_y : \alpha_y]_O = [b_1 : \beta_1, \ldots, b_n : \beta_n]_O \quad [a_x : \alpha_x, \ldots, a_y : \alpha_y] \in \binom{[a_1 : \alpha_1, \ldots, a_m : \alpha_m]_O}{[a_{i_1} : \alpha_{i_1}, \ldots, a_{i_m} : \alpha_{i_m}]_O}}{[a_1 : \alpha_1, \ldots, a_n : \alpha_n]_A = [b_1 : \beta_1, \ldots, b_n : \beta_n]_A} \text{EUNIANON}$$

$$\frac{\alpha_1 = \alpha_2 \quad \beta_1 = \beta_2}{\alpha_1 \to \beta_1 = \alpha_2 \to \beta_2} \text{EFUNC}$$

Fig. 3.: Rules for type-equivalence

For the records-types, the inference rules reflect the different outforms of a matching on names which we outlined in section 3 on naming in open distributed systems. For ERecEx all names of the record fields have to be exactly matching, that is they have to be syntactical equivalent. All types of the fields have to be pairwise identical. ERecOrd uses the structural information given by the position of a field as the matching criteria for names for an ordered record. All field names are renamed to their position and these renamed exact records have to be equivalent. Finally, ERecAnon discards the name-information completely. Two anonymous records are equivalent, if one can be permuted so that the permutation is equivalent to the other. The schema (...) denotes the set of permutations over a record.

EProd infers the equivalence of products from the equivalence on anonymous records. EUniEx, EUniOrd and EUniAnon define equivalence for exact, ordered and anonymous unions similar to those for records. As the last rule, EFunc defines equivalence of function types as the equivalence of argument and result types.

4.2 Rules for subtyping

The rules for subtyping infer judgments on a subtype-relation by interpreting the rules in figures 4 and 5 on an environment Γ which contains assumptions on subtype relations. SAss defines how these judgments are made.

Γ denotes a set $\{t_1 \mathrel{<\!\!\circ} s_1, \ldots, t_n \mathrel{<\!\!\circ} s_n\}$ of subtyping assumptions on type variables from which judgments on subtype relations are based (SAss). With rule SRefl, SAsym and STrans together with the minimal type $\bot$ – no value is of this type – (SMin) and $\top$ – all values are of this type – (SMax), $\mathrel{<\!\!\circ}$ is a partial order.

An exact record is a subtype of another, if the types of fields with identical names are in the subtype-relation (SRecEx). The subtype can have more fields, making it possible to substitute a value of the subtype for its supertype by forgetting the additional fields.

Rule SRecPos defines the subtype relation for ordered records. Here the names are replaced by their position in the record, thus using structural information for the matching on names. In contrast to the exact matching, additional fields can occur only at the end of the supertype.

Thus, the structural information "position" is too strong to what we intend. Therefore, we add rule SRecOrd allowing additional fields at any place of the subtype as long as the order of the fields also existent in the supertype is obeyed. The structural information we use here is "order" instead of "position" only.

Finally, SRecAnon discards names and structural information by defining subtyping on anonymous records by requiring one permutation of the subtype to be in the ordered subtype relation. As with the equivalence relation, subtyping on products is inferred from the subtype relation on anonymous records.

For unions to be in a subtype relation that ensures substitutability, the subtype may not have more variants than the supertype. SUniEx takes the syntactic name matching for exact unions into account. For ordered unions, SUniOrd

$$\frac{t \leqslant s \in \Gamma}{\Gamma \vdash t \leqslant s} \text{ SAss} \qquad \frac{\Gamma \vdash \alpha \leqslant \beta \quad \Gamma \vdash \beta \leqslant \alpha}{\alpha = \beta} \text{ SAsym}$$

$$\frac{\Gamma \vdash \alpha \leqslant \beta \quad \Gamma \vdash \beta \leqslant \gamma}{\Gamma \vdash \alpha \leqslant \gamma} \text{ STrans}$$

$$\frac{\alpha = \beta}{\Gamma \vdash \alpha \leqslant \beta} \text{ SRefl} \qquad \frac{}{\Gamma \vdash \bot \leqslant \alpha} \text{ SMin} \qquad \frac{}{\Gamma \vdash \alpha \leqslant \top} \text{ SMax}$$

$$\frac{\forall i \in \{1,\ldots,n\} : \Gamma \vdash \alpha_i \leqslant \beta_i \quad n \leq m}{\Gamma \vdash \langle a_1 : \alpha_1, \ldots, a_m : \alpha_m \rangle \leqslant \langle a_1 : \beta_1, \ldots, a_n : \beta_n \rangle} \text{ SRecEx}$$

$$\frac{\Gamma \vdash \langle 1 : \alpha_1, \ldots, m : \alpha_m \rangle \leqslant \langle 1 : \beta_1, \ldots, n : \beta_n \rangle \quad n \leq m}{\Gamma \vdash \langle a_1 : \alpha_1, \ldots, a_m : \alpha_m \rangle_O \leqslant \langle a_1 : \beta_1, \ldots, a_n : \beta_n \rangle_O} \text{ SRecPos}$$

$$\frac{\begin{array}{c}\Gamma \vdash \langle a_1 : \alpha_1, \ldots, a_i : \alpha_i \rangle_O \leqslant \langle b_1 : \beta_1, \ldots, b_j : \beta_j \rangle_O \\ j \leq i, j < l < n \\ \Gamma \vdash \langle a_k : \alpha_k, \ldots, a_m : \alpha_m \rangle_O \leqslant \langle b_l : \beta_l, \ldots, b_n : \beta_n \rangle_O \\ i < k < m, (n-l) \leq (m-k)\end{array}}{\begin{array}{c}\Gamma \vdash \langle a_1 : \alpha_1, \ldots, a_i : \alpha_i, \ldots, a_k : \alpha_k, \ldots, a_m : \alpha_m \rangle_O \leqslant \\ \langle b_1 : \beta_1, \ldots, b_j : \beta_j, b_k : \beta_k, \ldots, b_n : \beta_n \rangle_O\end{array}} \text{ SRecOrd}$$

$$\frac{\Gamma \vdash A \leqslant \langle b_1 : \beta_1, \ldots, b_n : \beta_n \rangle_O \quad A \in \begin{pmatrix} \langle a_1 : \alpha_1, \ldots, a_m : \alpha_m \rangle_O \\ \langle a_{i_1} : \alpha_{i_1}, \ldots, a_{i_m} : \alpha_{i_m} \rangle_O \end{pmatrix}, n \leq m}{\Gamma \vdash \langle a_1 : \alpha_1, \ldots, a_m : \alpha_m \rangle_A \leqslant \langle b_1 : \beta_1, \ldots, b_n : \beta_n \rangle_A} \text{ SRecAnon}$$

Fig. 4.: Rules for subtyping: partial order and records

discards the names but uses the positions of the variants. SUniAnon allows permutations to be used for anonymous unions.

For function types, we define a contravariant subtyping by rule SFunc. A function type is a subtype of another when its arguments are supertypes to those of its supertype and the results are subtypes to those of the supertype. By this, a subtyped function can safely replace its supertype function by discarding the additional arguments and by forgetting the additional results.

The type system we defined is underlying Laura's service type concept. The syntax and semantics of Laura's service type definition language are given in the next sections.

$$\frac{\Gamma \vdash \langle 1\!:\!\alpha_1,\ldots,n\!:\!\alpha_n\rangle_A \leqslant \langle 1\!:\!\beta_1,\ldots,n\!:\!\beta_n\rangle_A}{\Gamma \vdash \alpha_1 \times \ldots \times \alpha_n \leqslant \beta_1 \times \ldots \times \beta_n} \quad \text{SProd}$$

$$\frac{\forall i \in \{1,\ldots,n\} : \Gamma \vdash \alpha_i \leqslant \beta_i \quad n \leq m}{\Gamma \vdash [a_1\!:\!\alpha_1,\ldots,a_n\!:\!\alpha_n] \leqslant [a_1\!:\!\beta_1,\ldots,a_m\!:\!\beta_m]} \quad \text{SUniEx}$$

$$\frac{\Gamma \vdash [1\!:\!\alpha_1,\ldots,n\!:\!\alpha_n] \leqslant [1\!:\!\beta_1,\ldots,m\!:\!\beta_m] \quad n \leq m}{\Gamma \vdash [a_1\!:\!\alpha_1,\ldots,a_n\!:\!\alpha_n]_O \leqslant [a_1\!:\!\beta_1,\ldots,a_m\!:\!\beta_m]_O} \quad \text{SUniOrd}$$

$$\frac{\Gamma \vdash A \leqslant [b_1\!:\!\beta_1,\ldots,b_m\!:\!\beta_m]_O \quad A \in \binom{[a_1\!:\!\alpha_1,\ldots,a_n\!:\!\alpha_n]_O}{[a_{i_1}\!:\!\alpha_{i_1},\ldots,a_{i_n}\!:\!\alpha_{i_n}]_O}, n \leq m}{\Gamma \vdash [a_1\!:\!\alpha_1,\ldots,a_n\!:\!\alpha_n]_A \leqslant [b_1\!:\!\beta_1,\ldots,b_m\!:\!\beta_m]_A} \quad \text{SUniAnon}$$

$$\frac{\Gamma \vdash \alpha' \leqslant \alpha \quad \Gamma \vdash \beta \leqslant \beta'}{\Gamma \vdash \alpha \to \beta \leqslant \alpha' \to \beta'} \quad \text{SFunc}$$

Fig. 5.: Rules for subtyping: products, unions and functions

5 LAURA's service type definition language

A service type is required in all forms used in LAURA. A language called *STL* is defined to express a service type in a convenient way.

5.1 Syntax of LAURA's service type language

The abstract syntax of STL is defined in figure 6. To illustrate the syntax, we express the interface of a service offered or used by a traveling agency. It consists of three operations, getflightticket, getbusticket, and gettrainticket which take as arguments some identification of a credit-card, a travel date, and a destination. All operations confirm the booking and result in a price. getbusticket also results in the name of a bus-company. The type of this service is expressed in STL as in figure 7.

5.2 The semantics of LAURA's service-type definitions

The semantics of a type expression is defined as a type in the above type system. The notation $\Gamma(t)$ means the interpretation of t in the environment Γ.

Service-Type-Declaration ::=   (Signature-Declaration) **where**
    Type-Declaration* .
Signature-Declaration ::=   Operation-Signature*
Operation-Signature ::=   Operation-Name : Arguments $\rightarrow$ Results
Arguments ::=   type-name*
Results ::=   type-name*
Type-Declaration ::=   Type-name = Type-Definition
Type-Definition ::=   Predefined-Type |
    Type-Definition * ... * Type-Definition |
    ⟨ Type-Definition* ⟩ | [Type-Definition*]
Predefined-Type ::=   **string** | **character** | **number** | **boolean**

Fig. 6.: Abstract syntax of service type definitions language STL

```
(getflightticket: ccnumber * date * dest -> ack * price;
 getbusticket
: ccnumber * date * dest -> ack * price * line;
 gettrainticket : ccnumber * date * dest -> ack * price)
where
ccnumber = string;
date = <day,month,year>;
day = number;
month = number;
year = number;
dest = string;
ack = boolean;
line = string;
price = <number,number>.
```

Fig. 7.: A service type for travel booking in STL

Definition 1 Type expression semantics. For a term t from STL, the type denoted by t is written $\tau[\![t]\!]$ and defined as depicted in figure 8 with respect to the set of predefined types {string↦string, character↦character, number↦number, boolean↦boolean}.

The above example service type in figure 7 is interpreted in the environment

{string↦string, character↦character, number↦number, boolean↦boolean,
ccnumber↦string, date↦⟨number, number, number⟩$_A$, day↦number,
month↦number, year↦number, dest↦string, ack↦boolean, line↦string,
price↦⟨number,number⟩$_A$ }

τ[signature **where** type-declarations] =
    predefined $\cup$ τ[type-declarations](τ[signature])
τ[operation-signature$_1$; operation-signature$_2$] =
    $\langle \tau$[operation-signature$_1$], τ[operation-signature$_2$]$\rangle$
τ[operation-name : arguments –> results] =
    operation-name : τ[arguments] $\rightarrow$ τ[results]
τ[type-declaration$_1$; type-declaration$_2$] =
    τ[type-declaration$_1$] $\cup$ τ[type-declaration$_2$]
τ[type-name = type-definition] =
    {type-name $\mapsto \tau$[type-definition]}
τ[type-definition$_1$ * ... * type-definition$_n$] =
    $\langle \tau$[type-definition$_1$], ..., τ[type-definition$_n$]$\rangle_A$
τ[$\langle$ type-definition$_1$, ...,type-definition$_n$ $\rangle$] =
    $\langle \tau$[type-definition$_1$], ..., τ[type-definition$_n$]$\rangle_A$
τ[[type-definition$_1$, ..., type-definition$_n$]] =
    [τ[type-definition$_1$], ..., τ[type-definition$_n$]]$_A$
τ[**string**]   = string
τ[**character**]= character
τ[**number**]   = number
τ[**boolean**] = boolean

Fig. 8.: Semantics of STL-expressions

The interpretation results in the type

$\langle$getflightticket: $\langle$string, $\langle$number, number, number$\rangle_A$, string$\rangle_A \rightarrow$
        $\langle$boolean,$\langle$number,number$\rangle_A\rangle_A$,
    getbusticket: $\langle$string,$\langle$number,number,number$\rangle_A$, string$\rangle_A \rightarrow$
        $\langle$boolean,$\langle$number,number$\rangle_A$,string$\rangle_A$,
    gettrainticket: $\langle$string,$\langle$number, number, number$\rangle_A$, string$\rangle_A \rightarrow$
        $\langle$boolean,$\langle$number,number$\rangle_A\rangle_A\rangle$.

Let another booking-agency offer a service with the interface in figure 9. Interpreting this definition results in the type

$\langle$getflightticket: $\langle$string, $\langle$number, number, number$\rangle_A$, string$\rangle_A \rightarrow$
        $\langle$boolean,$\langle$number,number$\rangle_A\rangle_A$,
    getbusticket: $\langle$string,$\langle$number,number,number$\rangle_A$, string$\rangle_A \rightarrow$
        $\langle$boolean,$\langle$number,number$\rangle_A\rangle_A$,
    gettrainticket: $\langle$string,$\langle$number, number, number$\rangle_A$, string$\rangle_A \rightarrow$
        $\langle$boolean,$\langle$number,number$\rangle_A\rangle_A\rangle$.

Applying the subtyping rules on these types results in the judgement that the type defined in figure 7 is a subtype of the one in figure 9.

For LAURA then, when a serve-form contains the first service-description, it matches a service-form containing the second. The agent putting the serve-form into the service-space can then provide the service requested by the agent

```
(getflightticket: ccnumber * date * dest -> ack * cashed;
 getbusticket   : ccnumber * date * dest -> ack * cashed;
 gettrainticket : ccnumber * date * dest -> ack * cashed)
where
ccnumber = string;
date = <number,number,number>;
dest = string;
ack = boolean;
cashed = <number,number>.
```

Fig. 9.: A service type for travel booking in STL

that issued the service-form. Based on this mechanism, the LAURA-system takes appropriate coordination actions such as communicating the argument lists and as synchronizing the providing agent by returning a filled out serve-form.

6 Summary

In this paper we focussed on the issue of identifying services in open distributed systems. When using typed interfaces for this identification, the issue of how to deal with names in a type system becomes crucial. The assumption that a name uniquely represents exactly one meaning cannot be upheld in the light of open systems requirements.

We identified three outforms of matching on names that seem reasonable. Names can be used for a syntactical matching; some structural information such as position or order can be used for a matching, or names are discarded completely. We pointed out that all three outforms are qualitative equivalent with respect to a notion of soundness, as they all allow for unindented matches. Discarding names completely is the only outform that is complete in that is rules out unintended mismatches.

These considerations were taken into account for a type system with subtyping which includes these outforms in its formal rules. We applied the type system for the coordination language LAURA by finding a compromise between soundness and completeness.

In LAURA, services are identified by giving an interface consisting of operations offered. When deciding on the match of an offer to a service-request, we apply a syntactic matching on operation names but discard names completely for argument- and result-types of operations.

The application of the type system has been implemented as a recursive algorithm in a prototypical implementation of LAURA and has been defined for a parallel environment under coordination by the language ALICE ([Tolksdorf, 93a]).

While we have shown that the proposed outforms of dealing with names are equivalent under a qualitative evaluation, the quantitative effect still have to be analyzed. This required a statistical review of the usage of names in existing application of open distributed systems.

References

[Agha and Callsen, 93] G. Agha and C. Callsen. ActorSpaces: An Open Distributed Programming Paradigm. In *Proceeding of the Fourth ACM SIGPLAN Symposium on Principles and Practice of Parallel Programming*, 1993.

[Amadio and Cardelli, 91] Roberto M. Amadio and Luca Cardelli. Subtyping Recursive Types. In *Proceedings of the 18th annual ACM Symposium on Principles of Programming Languages*, pages 104–118, 1991.

[Cardelli, 88] Luca Cardelli. Structural Subtyping and the Notion of Power Type. In *Proceedings of the 15th annual ACM Symposium on Principles of Programming Languages*, pages 70–79, 1988.

[ISO/IEC JTC1/SC21, 93] ISO/IEC JTC1/SC21. Information Technology – Open Distributed Processing – ODP Trading Function, WG7 Working Document, 1993.

[ISO/IEC JTC1/SC21/WG7, 91] ISO/IEC JTC1/SC21/WG7. Basic Reference Model of Open Distributed Processing – Part 5: Architectural Semantics, Working Document, 1991.

[ISO/IEC JTC1/SC21/WG7, 93] ISO/IEC JTC1/SC21/WG7. Basic Reference Model of Open Distributed Processing – Part 1: Overview, Working Draft, 1993.

[ISO/IEC JTC1/SC21/WG7, 94a] ISO/IEC JTC1/SC21/WG7. Basic Reference Model of Open Distributed Processing – Part 2: Descriptive Model, Draft International Standard, 1994. ISO/IEC DIS 10746-2, ITU-T Draft Rec. X.902.

[ISO/IEC JTC1/SC21/WG7, 94b] ISO/IEC JTC1/SC21/WG7. Basic Reference Model of Open Distributed Processing – Part 3: Prescriptive Model, Draft International Standard, 1994. ISO/IEC DIS(E) 10746-3, ITU-T Rec. X.903.

[Raymond, 94] K.A. Raymond. Reference Model of Open Distributed Processing: a Tutorial. In J. de Meer, B. Mahr, and S. Storp, editors, *Proceedings of the International IFIP Conference on Open Distributed Processing*, pages 3–14. North-Holland, 1994.

[Tolksdorf, 93a] Robert Tolksdorf. Alice - Basic Model and Subtyping Agents. Technical Report 1993/7, Technische Universität Berlin, Fachbereich 20 Informatik, 1993.

[Tolksdorf, 93b] Robert Tolksdorf. Laura: A Coordination Language for Open Distributed Systems. In *Proceedings of the 13th IEEE International Conference on Distributed Computing Systems ICDCS 93*, pages 39–46, 1993.

[Tolksdorf, 94] Robert Tolksdorf. *Coordination in Open Distributed Systems (Preliminary title)*. PhD thesis, TU Berlin, 1994. Forthcoming.

An Efficient Implementation of Decoupled Communication in Distributed Environments

Andreas Polze
Freie Universität Berlin
Fachbereich Mathematik und Informatik
14195 Berlin, Takustr. 9
polze@inf.fu-berlin.de

ABSTRACT

Decoupled communication has proven to be a powerful paradigm for the development of applications in the context of open distributed environments. We have developed the *Object Space* approach which integrates that communication style with object-oriented techniques. It allows encapsulation of protocols for interactions in a distributed application into classes, thus providing a new level of abstraction.

Here we describe how decoupled communication as supported by the *Object Space* may be efficiently implemented in distributed environments. We start with a naive prototypical implementation and develop a more efficient fault tolerant solution. This implementation is itself distributed. It runs on top of UNIX systems connected by a LAN.

1. Introduction

The *Object Space* approach to distributed computation allows for decoupled communication between program components by providing a shared data space of objects. The *Object Space* approach extends the sequential language C++ with coordination and communication primitives as known from *Linda*. It integrates inheritance into the associative addressing scheme and facilitates passing of arbitrary objects between program components. Thus classes may be used to describe a protocol for communication between components of a distributed application.

Object Space itself is implemented in a distributed fashion. It employs several *Object Space Manager* processes running on different nodes of a network under UNIX. Firstly, we describe a naive prototype implementation of *Object Space*. Discussing how to avoid some performance drawbacks of this implementation, we develop an advanced solution. We use a technique called "optimistic asynchrony" to accelerate most *Object Space* operations. Furthermore we discuss how fault-tolerance through dynamic reconfiguration of communication links may be achieved.

In section 2 we briefly outline characteristics of the *Object Space* approach. Section 3 presents a prototypical implementation of *Object Space*. In section 4 we discuss a more efficient distributed implementation of *Object Space* in greater detail and describe how our implementation may be adapted for use in the context of very large distributed systems. In section 5 we give an overview of related work and finally, in section 6 we present our conclusions.

2. *Object Space* Operations

The *Object Space* approach for distributed programming [Polze 93a][Polze 93b] integrates a Linda-like communication style with object-oriented mechanisms such as inheritance, data encapsulation and polymorphism. *Object Space* constitutes a distributed associatively addressed memory. Components of a distributed application may access this memory and store or retrieve objects.

Four operations are available within *Object Space*:

- **out** writes an object into the *Object Space*. This operation works asynchronously.

- **in** and **rd** carry a template (a special object) as argument. Both operations retrieve a matching object from the *Object Space* and store its values in the template; they work synchronously and may eventually block until a matching object is found. **in** removes the object from *Object Space*. An object matches a template depending on its class and the values of its data components. If the template has less data components than a retrieved object (i.e., object's class is derived from template's class), extra components are silently discarded.

- **eval** creates a new UNIX process either locally or remotely. Either it carries the command line arguments for a remote process or it carries the address and arguments of a function which has to be executed locally as parameters.

Object Space relies on the abstractions "class" and "object". Objects serve as units of communication. C++ classes may be used for definition of communication protocols. These protocols describe the interactions between components of a distributed application. Previously defined protocols may be extended by use of inheritance. When dealing with objects in *Object Space* the C++ mechanisms of access control remain valid. Details of interaction may be hidden by construction of appropriate C++ classes. Those classes may provide secure interfaces to their clients. This view gives a new level of abstraction to the programmer of distributed applications.

The algorithm for associative addressing of objects within *Object Space* relies on an object's class and the values of its data components. A distributed type service is used to map classes' names onto unique integer identifiers. Inheritance may be expressed as a relation over those identifiers. Each pair "class name, identifier" is itself represented as a special object within *Object Space*.

3. A prototypical implementation

The *Object Space* approach supports the idea of a distributed shared memory. Our prototype implementation in C++ is based on interprocess communication mechanisms available within the UNIX operating system.

A scenario around *Object Space* as shown in Fig. 1 includes several processes, each of them storing objects into and retrieving them from *Object Space*. Some kind of storage medium is needed to keep the data which makes up an object available. In our implementation these storage media are provided by a separate UNIX process, the *Master Object Space Manager*. It uses virtual memory to store the objects. Furthermore this process performs matching between templates and objects. In addition to the *Master Object Space Manager* several *Client Object Space Manager* processes may exist. These processes do not store any object at all. They just forward operation requests to the *Master Object Space Manager* process using a special protocol on top of TCP/IP. Thus crossing machine boundaries is possible.

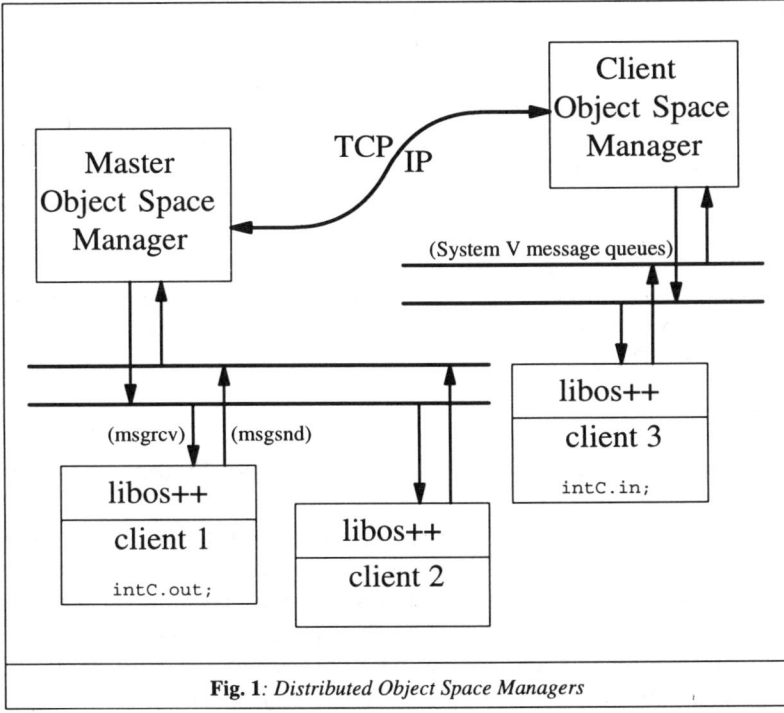

Fig. 1: *Distributed Object Space Managers*

Client processes communicate with a local *Object Space Manager* process. This process may be either the *Master Object Space Manager* or a *Client Object Space Manager* as shown in Fig. 1 . They use message queues as the communication mechanism. Message queues have been introduced with UNIX System V but actually they are available with nearly every UNIX System. A library libos++ transforms calls to *Object Space* operations into a proper sequence of msgsnd()

and `msgrcv()` system calls on these message queues. Client programs may access the operations from `libos++` trough inheritance from two communication base classes. An *Object Space Manager* handles two message queues, a read queue and a write queue. Each client process contacts the read queue of a local *Object Space Manager* to initiate operations on *Object Space*. Results of operations are returned on the corresponding write queue.

The operation **out** initiates a write-operation on the read-queue of a local *Object Space Manager*. It transmits an *actual* object and does not return a result. On the other hand, the operations **rd** and **in**, called on client side, write a template object onto the read-queue of a local *Object Space Manager* and block until it returns a matching object. The matching object is returned on the corresponding write-queue. Both read- and write-queue are multiplexed, thus allowing any number of clients to communicate with the same *Object Space Manager* at once.

Matching between templates and *actual* objects is performed by the *Master Object Space Manager*, so this process is a potential performance bottleneck. But we discover another performance problem when looking at communication delays. Practical experience has shown that communication via message queues is much faster than via a TCP/IP connection. We have measured that one *Object Space Operation* takes roughly 1.05 ms on a DECstation 5000/125 if communication occurs over message queues only. It takes about 10 ms if local TCP/IP communication (which is entirely handled in software) is involved. And finally, the same *Object Space Operation* issued on a remote site, performed using message queues and TCP/IP, takes about 21 ms. We have obtained similar results on Sparc 2. On a 386based PC-Unix the operations took about twice the time. So our prototype implementation of *Object Space* is acceptable on multiprocessor systems where all communication can be performed via message queues. For a real distributed environment our implementation seems to be suitable only if *Object Space* operations are performed infrequently.

4. A more efficient implementation of *Object Space*

The *Master Object Space Manager* plays an integral role within the prototypical implementation of *Object Space* as described in section 3. It maintains all the objects contained in *Object Space* and performs the matching between templates and actual objects. As each call to an *Object Space* operation has to pass this process, it may soon become a performance bottleneck. To deal with this problem we replicate data. Objects are stored within several *Object Space Manager* processes. So each of the synchronous operations **in** and **rd** has to access local data only. Fast communication via message queues is used in that case. Another problem is ensuring consistency. Although two copies of a certain object may exist in different *Object Space Managers*, only one client performing an **in** operation must be able to obtain this object. To achieve this we need a special protocol for communication between *Object Space Managers*.

In general, replication of objects in several *Object Space Managers* is useful only if those objects are accessed from different nodes in the network at nearly the same time. To implement associative addressing in the *Object Space*, we have employed a distributed type service. Prior to each *Object Space* operation concerning an instance of class "x", the type identifier for that class has to be obtained. At least the first creation of an instance of class "x" in a program component includes communication between the particular program component and an *Object Space Manager*. Thus, an *Object Space Manager* is able to know the classes of objects which are used by its local clients (components of a distributed application). Employing this knowledge, a rule for replication of objects on different *Object Space Managers* may be defined. This rule has to consider the associative addressing scheme and inheritance. It consists of two parts:

1) "A copy of each object of class 'x' contained within *Object Space* has to be stored on a node if at least one *Object Space* operation using an instance of class 'x' as argument was performed on this node."

2) "A copy of each object of class 'y' contained within *Object Space* has to be stored on a node if at least one operation using an instance of a class 'x' which is the ultimative base class of 'y' was performed on this node."

When formulating this rule we assume that a node is represented by an *Object Space Manager* process. In a first attempt we assume that communication channels are established between *Object Space Manager* processes on different nodes to form a fully connected graph.

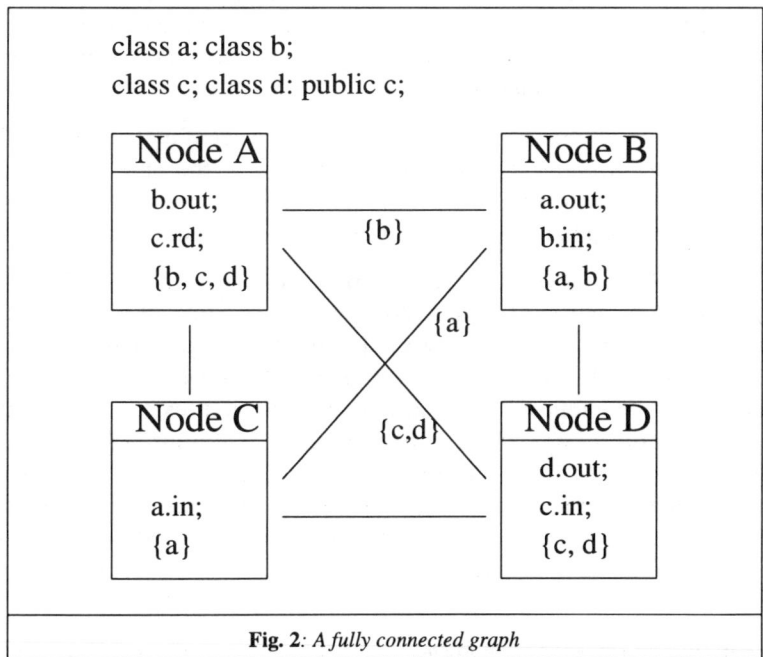

Fig. 2: *A fully connected graph*

In Fig. 2 four processes on different sites called Node A trough Node D are fully connected via TCP/IP communication links. Those processes implement the *Object Space*. They store instances of classes a, b, c and d. Class d is derived from class c. In Fig. 2 a set of classes is associated with each node. These sets describe classes whose instances are stored on a node according to the replication rule formulated earlier. The *Object Space* operations printed within the node's boxes in Fig. 2 are performed by clients of the corresponding *Object Space Managers*. For simplicity we have omitted the clients of *Object Space* from our drawing. Instead of declaring variables of the appropriate classes we have used the classes' names to indicate *Object Space* operations here.

One may notice that Node A stores instances of class d, although no *Object Space* operation concerning this class has happened on Node A. But due to the associative addressing scheme used within *Object Space* the operation c.rd may eventually return an object of class d. Communication links between *Object Space Manager* processes are used to run a protocol to ensure consistency among replicated objects. In Fig. 2 the links are labeled showing the classes whose instances are transmitted via a particular link.

Now we want to discuss in greater detail how the *Object Space* operations **out** and **in** work in context of our replication scheme. An unique object identifier and a replication counter are assigned to each object and its replicas within *Object Space*. This identifier is composed from the node name where the object initially appeared (via operation **out**) and a serial number. Between *Object Space Manager*s, our algorithm uses message passing via TCP/IP to distribute an object's copies. Operations **out** and **in** employ message passing between remote *Object Space Managers* whereas operation **rd** works locally. Four different types of messages are used.

Replication of objects may be achieved by transmitting the object's data to all *Object Space Manager*s which store instances of the object's class. Thus *Object Space* operation **out** results in broadcasting of messages of type "out". These messages contain the object's identifier, its replication counter and its data.

The implementation of operation **in** is more sophisticated. Three different message types are used by our distributed realization of operation **in**. A message of type "initiate-in" expresses the request of a particular *Object Space Manager* to obtain all copies of an object. A remote *Object Space Manager* may answer with a "commit-in" message. Eventually it tries to perform an **in**-operation on another copy of the same object. Then it refuses the request from the *Object Space Manager* mentioned first and sends a "retry-in" message. Eventually an **in**-operation needs several iterations of message passing between holders of replicas of a particular object. A heuristic is used to serve conflicts between competing *Object Space Manager*s when accessing copies of the same object.

Firstly, the *Object Space Manager* performing an **in** generates a sequence of pairs (seqno,priority) where priority is a random number. A priority is attached to each message of type "initiate-in" and "retry-in". If several *Object Space Manager*s compete for a particular object, the priority of a "initiate-in" message decides which of them succeeds. An *Object Space Manager* receiving an "initiate-in" message has four principal patterns of reaction:

1) It is not interested in the particular object. In that case no client of the *Object Space Manager* attempts to access the particular object via operation **in**. The *Object Space Manager* sends a "commit-in" message and deletes its copy of our object.

2) It has sent an "initiate-in" message concerning the requested object with lower priority as the received message. In that case the *Object Space Manager* also sends a "commit-in" message and deletes its copy of our object.

3) It has sent an "initiate-in" message concerning the requested object with the same priority as the received message. Then it increments the sequence number by one and sends a "retry-in" message back.

4) It has sent an "initiate-in" message concerning the requested object with higher priority. Then it silently waits for further messages.

When receiving a "commit-in" message an *Object Space Manager* decrements the replication counter of the object mentioned within the message. If that counter approaches one, the *Object Space Manager* holds the last copy of our object within the system. It safely may forward the object to its client. In Fig. 3 three *Object Space Manager* processes exchanging messages are shown. Node A and Node C compete for a particular object. Messages "initiate-in", "commit-in" and "retry-in" are denoted (i), (c) and (r), respectively. Numbers in parenthesis describe the priority of the corresponding messages.

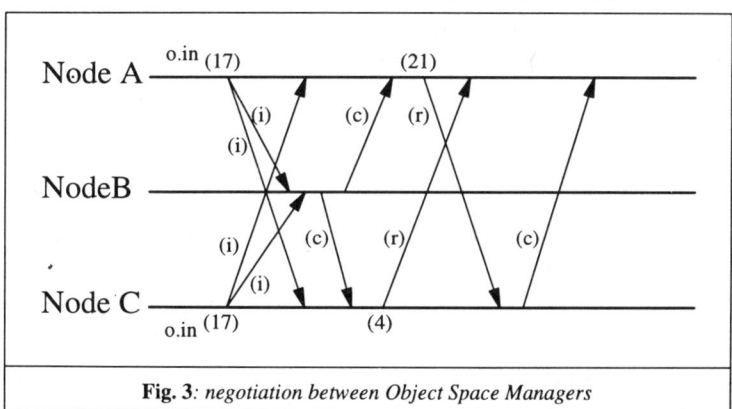

Fig. 3: *negotiation between Object Space Managers*

If an *Object Space Manager* receives a "retry-in" message it chooses another random number as priority and sends a new "initiate-in" message. The newly generated random number is stored under the appropriate sequence number. If a "retry-in" message appears with a sequence number lower than the current one, the previously stored priority for that sequence number is used for the next "initiate-in" message.

The number of iterations needed for an **in**-operation depends on size and distribution of the random numbers used to denote the priority of an "initiate-in" message. However, if the number of *Object Space Manager* processes is small compared with the maximum random number, the likelihood of performing an **in**-operation in a single iteration is high.

Using the implementation scheme described above, we can now attempt to figure out costs (i.e. communication delays) of *Object Space* operations.

- From a client's view, operation **out** may be handled asynchronously. A client simply stores an object on its local *Object Space Manager* and may continue execution. This operation involves message queue communication only and therefore has very little delay. The *Object Space Manager* distributes the new object via the communication links labeled with the object's class name and attaches object identifier and replication counter with the object.

- The operation **rd** is a synchronous operation. Due to our replication scheme this operation may act on data stored in the local *Object Space Manager*. Thus only fast message queue communication needs to be used. Because operation **rd** has no impact on the contents of *Object Space*, no communication between *Object Space Manager* processes is needed to perform that operation. Operation **rd** may eventually block. In that case the unsatisfied operation request is stored in the local *Object Space Manager*.

- The operation **in** also works synchronously. A client process has to wait until the local *Object Space Manager* returns a matching object. Before returning an object, the *Object Space Manager* has to request that particular object from all nodes holding copies. This is managed using the algorithm described above. But unfortunately, slow TCP/IP communication has to be used by the *Object Space Managers* to negotiate about a particular object. In the case of a blocking **in** operation, i.e. the local *Object Space Manager* has no matching object available, the unsatisfied operation request is stored.

With our new implementation scheme we have shown, how *Object Space* operations **out** and **rd** may be realized with little delay using fast message queue communication and locally stored replicas of objects. Now we want to discuss how the performance of operation **in** can be further improved.

When performing an **in** operation an *Object Space Manager* chooses a matching object, forwards it to the client and deletes it from the *Object Space*. Since the object may be replicated, all holders of replicas have to commit to the above steps and have to delete their copies of the object too. Our replication scheme guarantees that as few replicas as possible are stored within the system. However, the communication between *Object Space Managers* increases the time an **in** operation takes. Perhaps another *Object Space Manager* may try to access the object at the same time. Then a heuristic determines which of them succeeds. Eventually our *Object Space Manager* has to choose another matching object and to negotiate again.

Instead of waiting for commitments from all holders of replicas of a particular object, we may forward the object to the client process early. Then the client process may continue execution. Eventually this may be wrong since the *Object Space Manager* may fail when negotiating with the other holders of an object's copies. So the client process must be able to set back to the point of execution where it received the object from *Object Space*. Then it must be able to continue execution with another object. A transaction-like mechanism is needed to implement this behaviour. Our approach softens the semantics of operation **in**. We call this technique "optimistic asynchrony" since the result of an otherwise synchronous operation is used before the operation is entirely completed.

To ensure consistency within *Object Space* a client returning from an **in** operation gets blocked when performing the next *Object Space* operation until all *Object Space Managers* have committed about deletion of the **in** operation's object from *Object Space*. Nevertheless we are able to perform **in** operations with the delay of message queue communication only if a client does not perform sequences of *Object Space* operations, an assumption which is true in many cases.

Setting breakpoints during the execution of a client's process and returning to such a breakpoint can be implemented by different means. The UNIX system provides C library routines `setjmp`/`longjmp` which allow to restore the stack of a process. An extra effort has to be made to keep a process' heap consistent when performing a call to `longjmp`. In [Bilris at al. 93] the authors discuss how overloading of operator new in a C++ class may be used to keep track of a process' heap. The technique described there is used to implement persistent objects in a database system. Overloading of operator new for all classes whose instances appear as arguments to *Object Space* operations allows to take care of the heap management when calling `longjmp`. To implement a transaction scheme we additionally need an inverse operation for each function which uses the result of an **in**-operation. From a client's point of view, once these operations have been specified, setting a process back to a breakpoint may happen fully transparently.

Another way to implement "optimistic asynchrony" would be to keep an invisible reference to the object returned from operation **in** in the library `libos++`. Later on, if negotiation about the object retrieved from *Object Space* has failed, this reference may be used to modify the previously retrieved object. The client process has to support a notification function which is called by the library `libos++` in such a case. This technique does not work transparently, the programmer of a client process has to deal with a "failed" operation **in** explicitly. However, the notification function is somewhat similar to an inverse operation in the transaction scheme mentioned earlier. But it allows greater flexibility. For example, instead of un-doing all the operations on a particular object the programmer can decide to simply delete the object. To avoid the extra effort introduced by dealing with a notification function, a client process' programmer may enforce operation **in** to work synchronously. Thus the original semantics of that operation may be obtained.

We have presented an implementation scheme for the *Object Space* approach which allows performance of *Object Space* operations with basically the costs of local interprocess communication even in distributed environments. To achieve this result we employ replication of objects stored in *Object Space* and "optimistic asynchrony" of **in** operations. Our replication scheme guarantees as few replicas as possible of a particular object within the system. Additionally, with our new implementation scheme we are able to address another topic: fault tolerance.

Fault tolerance

In open distributed environments components of a system may disconnect from the system and later reconnect without being considered faulty. The *Object Space* approach should be able to deal with those cases. Although the breakdown of a network connection may be due to a scheduled downtime and thus be intended, we consider it as a fault. Our implementation of *Object Space* is able to respond to certain faults through reconfiguration.

We consider two kinds of failures: the breakdown of a network connection and the breakdown of a node, represented by an *Object Space Manager* process. Since we use TCP/IP as the communication mechanism, both cases can be easily detected by a surviving *Object Space Manager.* In those cases the UNIX system call read() returns a value of 0. When receiving such a result a process tries to establish a connection to the initial communication partner via a third node. This succeeds if the initial partner survived and results in a reconfiguration as shown in Fig. 4 .

In our example shown in Fig. 4 the direct connection between Node B and Node C is assumed to be faulty. All communication regarding operations **out** and **in** for objects of class a now has to be performed via Node D. This node acts as a bridge. The functionality of *Object Space* is not affected by that reconfiguration.

In general, if a node cannot reach a particular communication partner it subsequently asks its remaining partners for an connection to that particular node. If none of them succeeds the node is considered to be faulty. In our example, if Node C cannot reach Node B, neither via Node D nor via Node A, then Node B is considered to be faulty (or disconnected). In that case the replication counter attached with those objects for which Node B held a copy has to be decremented. Our fully connected graph with four nodes from Fig. 4

degenerates in that case into a fully connected graph with three nodes and a single node. Clients of *Object Space* using only Node A, Node C and Node D can proceed as usual. On the other hand, if we assume that Node B is simply disconnected for some reason, clients of Node B can proceed too.

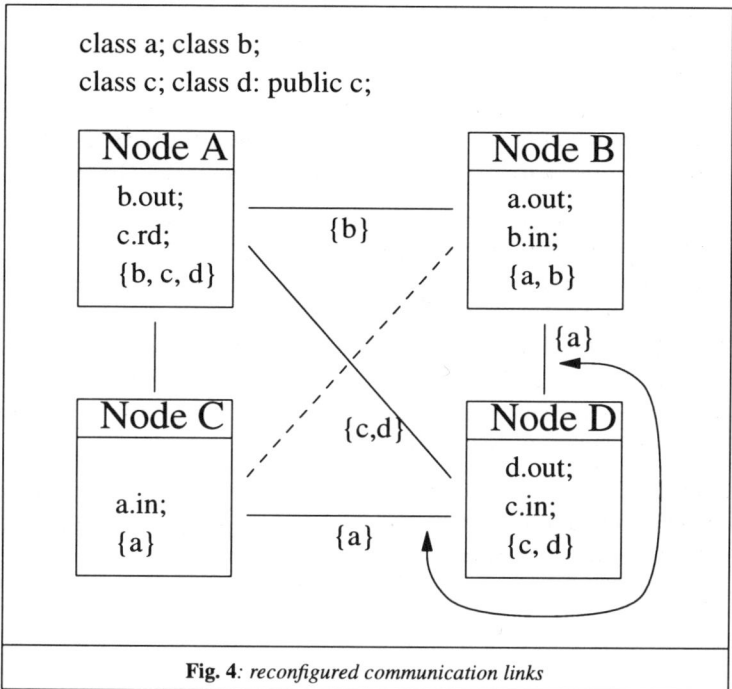

Fig. 4: *reconfigured communication links*

An *Object Space Manager* periodically attempts to reconnect to each of its former communication partners. Within our example, the connection from Node B to another node may become re-established. Then this node would act as a bridge. Objects with the same identifier stored on Node B and on any other node would be considered to be replicas. Then only the replication counter attached to the replicas of such an object would have to be adjusted. Objects of class b which are stored on Node B but not on Node A would be replicated onto Node A and vice versa.

We have shown, how our new implementation scheme of *Object Space* can deal with network failures. Furthermore, we have described a method whereby the temporal disconnection of a node may be handled by the *Object Space*. Both cases are common in open distributed systems. Thus the predicate "fault tolerant" describes the usability of our approach in the context of open distributed environments.

Other interconnection networks

Our new implementation scheme for the *Object Space* bases on a fully connected graph. *Object Space Manager* processes represent nodes within that graph. Considering a set of n nodes, each node has to manage n-1 communication links. This may impose some restrictions on the size of systems using *Object Space*. For example, under the Berkeley derivate of the UNIX operating system each process can maintain 64 TCP/IP connections simultaneously. So we can connect at most 64 *Object Space Manager* processes. Other environments are worse: a transputer supports 4 communication links.

To overcome this problem we can introduce additional processes which are tightly coupled with an *Object Space Manager*. These processes would act as bridges and enlarge the number of communication links per node. At least on a UNIX system, communication between the *Object Space Manager* process and a bridge node can be performed again via fast message queues. If we now define a node as consisting of an *Object Space Manager* and several "bridge" processes, we can use the same simple routing scheme for communication between nodes as described earlier.

Another way to deal with a very restricted number of communication links would be to use a different network topology for interconnection between *Object Space Managers*.

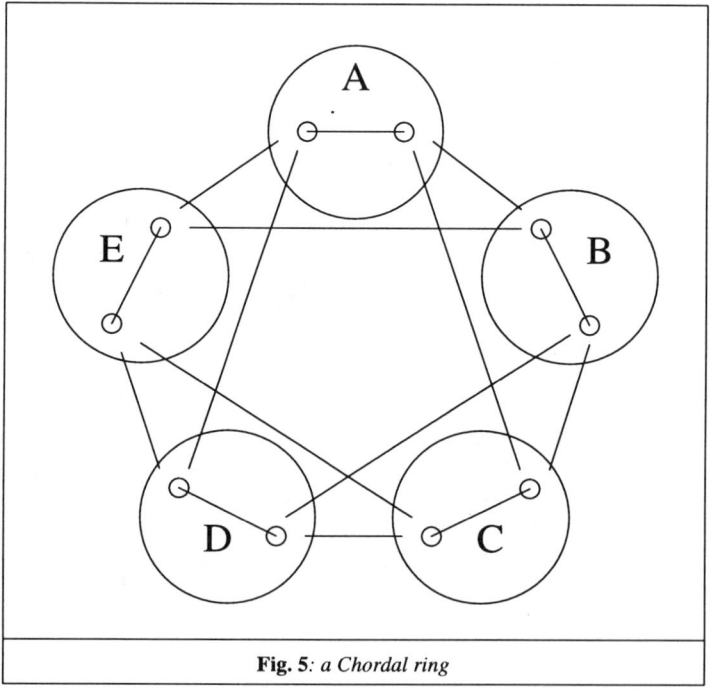

Fig. 5: *a Chordal ring*

We allow the maximum elementary path between two nodes to have a length d. Furthermore a node should have an in-out degree (number of communication links) of k. Then a Moore graph [Lipovski/Malek 87] is defined by the property that the number of nodes approaches the Moore bound for given d and k, which is

$$\frac{(d^k(d-1)^k - 2)}{d-2}$$

Moore graphs have an attractive property: the number of arcs is proportional to the number of nodes, since k is fixed, and the path length between any two nodes is bounded by d. Thus a Moore graph provides the most number of nodes for a given d. In Fig. 5 we show a Chordal ring, which is a Moore graph with k = 3 and d = 3.

Within Fig. 5 small circles describe communication endpoints. We assume that each of them is implemented by a separate UNIX process. We can combine pairs of those processes into *Object Space Managers* as indicated by the big circles in Fig. 5 . If we consider these big circles as nodes in our network we again have a fully connected graph and can employ the routing scheme described earlier using links labeled with sets of classes' names for communication between *Object Space Managers*. Although the number of communication links per UNIX process is bound by 3 in our example each *Object Space Manager* as indicated by a big circle in Fig. 5 has more than 3 network connections. Thus an interconnection scheme based on a Chordal ring is suitable for the implementation of *Object Space* in very large networks.

Another view of Fig. 5 would consider each of the small circles to be a separate *Object Space Manager* process. This would require development of a new routing scheme for communication between those processes. However, since each of them uses k communication links only, for k = 4 we could easily tailor an implementation of *Object Space* which runs in a transputer environment.

Unfortunately, Moore graphs are defined by their ability to meet the Moore bound, rather than by construction or axiom. So generally we cannot get a Moore graph of given size (d, k). If we do get it, we generally do not have a routing and scheduling algorithm for it.

5. Related Work

The *generative communication* style was first proposed within the context of *Linda* [Gelernter 85]. *Linda* was developed for programming parallel computers and therefore has a slightly different focus than approaches intended for programming distributed systems. However, as well as a parallel implementation in [Bjornson et al. 1987] a distributed implementation of

Linda is also described. This implementation makes use of various preprocessing techniques embedded within the *Linda* precompiler. The precompiler analyzes a parallel program with all its components as a whole, a technique which is not usable in the context of open distributed systems. A further description of precompiler techniques may be found in [Carriero/ Gelernter 91].

One drawback of *Linda* is that its communication constructs are rather low-level. This issue is addressed in [Ahmed 91]. The use of object-oriented techniques to extend *Linda* with higher levels of abstraction has been the major idea behind *Object Space*. Other approaches which combine *Linda*-like communication and object-orientation may be found in [Jellinghaus 90] and [Matsuoka 88]. However, in these papers no issues regarding a real distributed implementation are mentioned.

In [Callsen et al. 91] two distributed implementations of a *Linda*-like system are described. One of them runs under the UNIX operating system, the other under Helios on top of transputers. The transputer version uses a ring topology for communication between nodes storing tuples. Within that system, requests for tuple space operations travel along the ring from node to node, eventually matching against tuples already stored on a particular node.

6. Conclusions

We have described two distributed implementations of *Object Space*. The first, a naive prototypical implementation, shows some performance drawbacks due to a centralized *Master Object Space Manager* process and communication delays caused by TCP/IP. However, this implementation seems to be suitable for multiprocessor systems running UNIX where all interprocess communication can be handled via message queues.

In the second, more efficient implementation *Object Space* is realized as a fully connected graph of *Object Space Manager* processes. This implementation introduces a replication scheme for objects. Together with "optimistic asynchrony" of **in** operations, this technique allows most *Object Space* operations to be performed without delays caused by communication with a remote site.

In the context of our new efficient implementation we have discussed how network failures and breakdown of nodes in a distributed environment may be handled. We have proposed an algorithm whereby our new implementation of *Object Space* may deal with those cases.

Investigating other topologies for interconnection of *Object Space Manager* processes we have briefly discussed how our implementation of *Object Space* may be adapted for very large distributed systems.

Furthermore we have shown how an implementation of *Object Space* for a transputer environment may be realized.

References

[Ahmed 91] S.Ahmed, D.Gelernter; *Program Builders as Alternatives to High-Level Languages*; Report YALE/DCS/RR-887, Yale University, Dept. of CS, November 1991.

[Bilris at al. 93] A.Bilris, S.Dar, N.H.Gehani; *Making C++ Objects Persistent: the Hidden Pointers*; Software-Practice and Experience, Vol. 23(12), December 1993.

[Bjornson et al. 1987] R.Bjornson, N.Carriero, D.Gelernter, J.Leichter; *Linda, the Portable Parallel*; Research Report YALE/DCS/RR-520, February 1987.

[Callsen et al. 91] C.J.Callsen, I.Cheng, P.L.Hagen; *The AUC C++ Linda System*; in Technical Report 91-13, Edinburgh Parallel Computing Centre, Greg Wilson (Editor).

[Carriero/Gelernter 91] N.Carriero, D.Gelernter; *New Optimization Strategies for the Linda Pre-Compiler*; in Technical Report 91-13, Edinburgh Parallel Computing Centre, Greg Wilson (Editor).

[Gelernter 85] D.Gelernter; *Generative communication in Linda*; ACM Transactions on Programming Languages and Systems, 7(1):80-112, 1985.

[Jellinghaus 90] R.Jellinghaus; *Eiffel Linda: An Object-Oriented Linda Dialect*; ACM Sigplan Notices, Vol.25, No.3, December 1990.

[Lipovski/Malek 87] G.J.Lipovski, M.Malek; *Parallel Computing — Theory and Comparisons*; John Wiley & Sons, Inc., 1987.

[Matsuoka/Kawai 88] S.Matsuoka, S.Kawai; *Using Tuple Space Communication in Distributed Object-Oriented Languages*; Proceedings of Conference on Object-Oriented Programming Systems, Languages and Applications (OOPSLA) '88.

[Polze 93a] A.Polze; *The Object Space Approach: Decoupled Communication in C++*; Proceedings of Technology of Object-Oriented Languages and Systems (TOOLS) USA'93, Santa Barbara, August 1993.

[Polze 93b] A.Polze; *Using the Object Space: A Distributed Parallel make*; Proceedings of 4th IEEE Workshop on Future Trends of Distributed Computing Systems, Lisbon, September 1993.

[Polze 94a] A.Polze; *Interactions in Distributed Programs based on Decoupled Communications* to appear in Proceedings of Technology of Object-Oriented Languages and Systems (TOOLS) USA'94, Santa Barbara, August 1994.

Extending the Rôle of Object References in Distributed Systems

Peter Dickman

Department of Computing Science, University of Glasgow, Glasgow G12 8QQ, UK

Abstract. Distributed systems make heavy use of references to objects, which provide a location mechanism as well as potentially supporting location and access transparency. Sophisticated implementations of references may have integrated garbage collection, or support usage monitoring for load balancing and form a key component of replica management systems. This paper has two main parts: the current uses and implementation techniques for references are reviewed; and possible future uses of references are discussed. In the latter part the proposed uses of extended references in the DRASTIC project are presented. The DRASTIC project addresses type evolution, integration of legacy systems, and large heterogeneous distributed object systems in a unified architecture incorporating an additional level of structuring (zones), and extended references between zones (change absorbers).

On Protocols for Loss-less Statistical Multiplexing in Integrated Networks

Mihai Mateescu[1]

GMD FOKUS Berlin
Hardenbergplatz 2, D-10623 Berlin, Germany
email: mateescu@fokus.gmd.de

Abstract. The paper analyses the options available for bandwidth management in ATM-based networks carrying predominantly data traffic, develops a rationale for loss-less statistical multiplexing and quantifies the trade-offs associated with these approaches. Distributed computing, multimedia mail, walk through virtual reality are just a few examples of applications where the current approaches for traffic control (as foreseen by ITU-TSS) are difficult to use. Very high speed networks supporting integrated services will evolve towards new, data-like resource management approaches. The new protocols allow for a good utilization of the links without preliminary knowledge of the source behaviour.

1 Introduction

Traffic and congestion control has been recognized as an essential factor in supporting in an efficient way both the connectionless and the connection oriented services provided in broadband integrated networks. There are basically two options which have been envisaged for bandwidth management. The first approach has been to analyse the possibility of using the extensive theoretical insight gained in circuit switched networks for the bandwidth management of the new, ATM-based BISDN. The idea is to use a measure of the statistically multiplexed connections, based on which a new call can be admitted in the network or not, according to a space of admissible states. This measure, also called equivalent bandwidth, requires preliminary knowledge of the source behaviour and the possibility to express this behaviour in terms of a tractable set of parameters. The analysis of the bandwidth allocation schemes shows that tractable expressions for the equivalent bandwidth [GAN_91] can only be obtained under simplifying assumptions (such as normal distributions or Poisson processes). A solution which is not so restrictive [Hui_88], [Kel_91] can be obtained using large deviation theory, but one restriction applies: the source process must be stationary. Since traffic measurements made public in 1993 suggested that data traffic actually behaves like a growth process (fractal property), it became clear that traffic control decisions based on the equivalent bandwidth would not be suitable for data traffic. As

---

1. This work was supported by the Commission of the European Communities under the Research for Advance Communications in Europe - RACE II Project 2068

a result, the second approach [RACE_94] concentrated on solutions where no bandwidth is reserved for a particular connection, and the source can transmit on a best effort basis, grabbing as much resources as available. When resources are not available, a congestion control scheme is activated to avoid any losses (loss-free). This approach does not require a preliminary knowledge of the source behaviour, nor is the stationarity of the source process an issue for guaranteeing that the mechanism is loss-free. The trade off is that there are no guarantees for the access delays.

2   Traffic Management Using the Reservation Approach

The management of the concurrent access to the network and link resources in B-ISDN has been structured according to the cell-burst-call time-scales, in an effort to extend the allocation principles developed for circuit-switched networks. Each control level is responsible of the next lower level, and it should attempt to provide a guarantee for the layer at the smaller time-scale so that the latter can meet its QOS (especially in the form of an acceptable loss rate) requirements for all services. This has been stated in [Hui_88] as the principle of layered switching. The hope is that it will be possible:

> (a) To declare at connection set up time a user-provided descriptor of the activity of the source
> (b) To derive from this descriptor a statistical measure of the resource requirements (bandwidth, buffers) of the mix of sources admitted in the network, and to decide based on it whether to accept a new connection, very much like currently done in circuit-switched networks
> (c) To police the traffic pattern offered by the bursty sources, so that it conforms with the parameters declared at step (a).

3   A Critical View of the Reservation Approach

Most of the insights we have gained in the area of network dimensioning are based on the realities of circuit switched networks carrying a majority of voice traffic. Today voice calls make up to 85% of the revenues in the large public networks and this proportion will not change dramatically in the near future. A large amount of knowledge about the behaviour of these networks has been obtained, mainly based on the assumptions of negative exponential distributions for the characterization of the arrival processes, which leads to very tractable analytical approaches and to closed forms. *These assumptions are not only tractable, they also correctly model the reality, since the speaker process and many other natural processes (such as population growth, etc.) are indeed exponential.* But the data and data-like traffic streams generated from file-transfers, video codecs, digital measurement instruments are no longer "natural" processes, they are the result of an important number of processing stages: coding, gaining access to bus cycles, software loops, all of which modify the statistics of the original transactions, even if the original transactions were generated with an exponential rate. It is a quite acceptable assumption that the interarrival times for human requests for file transfers are exponentially distributed, but the processing of the request (reading the file from the disk, transferring it over the bus to the memory and then to the network controller) requires a "thinking time" and *the interdeparture times will have distributions dependent on the service type.*

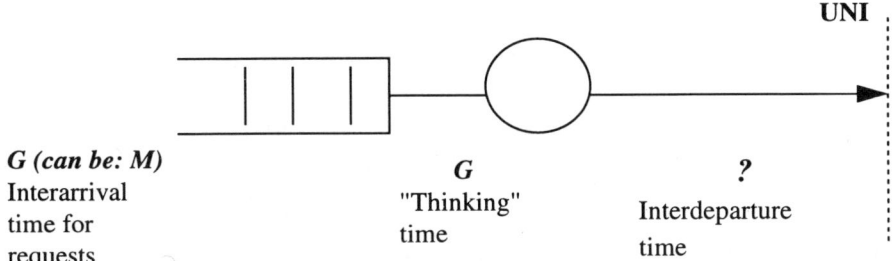

G (can be: M)
Interarrival
time for
requests

G
"Thinking"
time

?
Interdeparture
time

UNI

Most of the results obtained for the reservation approach are based on the assumption of exponentially distributed burst and silence periods, *but the interdeparture times of the bursts at the UNI will not even be a renewal process in the general case.*

This intuitive reasoning is backed by measurements made on recorded real traffic on an inter-LAN multiplexer reported in the remarkable papers by Fowler and Leland [FL_91] and Leland et. al. [LTWW_93]. Among other interesting observations, two are of special interest for this report: the *observed traffic characteristics* and *the behaviour during congestion.* According to [FL_91] and [LTWW_93],the fundamental feature of the traffic is its burstiness at every time scale, from milliseconds to months. In other words, no time scale exists where the random variation of the process would appear to be stationary, and this can be quantified using the index of dispersion of arrivals (the ratio between the variance and the mean of the arrival rate). In this case, the index of dispersion of arrivals is logarithmically linear *(fractal property)* as opposed to stationary processes where it is a constant (for Poisson to unity). Congestion behaviour has similar patterns at very different time scales and is typically preceded by a rise in traffic intensity. Congestion duration is not dependent on the network load, an increased load doesn't lead to longer congestion periods, but to more frequent short congestion periods. *These observations make the reservation approach and the underlying "principle of layered switching" unsuitable for allocating resources for data and data-like traffic.* The idea behind this principle was that the bandwidth requirement at the call level could be forecasted based on a statistical measure of the resource requirement of the processes at the burst level. The slower call level would appear to be stationary compared to the millisecond time scales of the burst layer. But the new facts outlined above show that by no means could the burstiness be narrowed down to a single time scale. This makes the dimensioning of the network based on the current approaches no longer applicable. There are two solutions to this problem:

(i) To develop new traffic models, capable of capturing the *fractal property* mentioned above. One such candidate model are the so called *self-similar processes*, addressed in the references [FL_91], [LTWW_93].

(ii) To use (where possible) bandwidth management procedures which do not use advance reservation and as such *do not rely on a statistical measure of the bandwidth requirements* (and on some assumptions about the underlying processes), but rather react dynamically to changes of the network load, by restricting access to the network or to a certain area of the network. This approach does not use traffic descriptors. The dimensioning of the network can be made using long term load statistics.

4   A Rationale for Loss-less Statistical Multiplexing

The emphasis of resource management in very high speed integrated networks is shifted to the management of the inherent concurrent access to and from the independent parts of a complex distributed system. The concurrent nature of the resource sharing in traditional telephone networks is hidden through the process of connection establishment, which effectively keeps the competition for resources outside the network.The media access protocols on the other hand have been used in traditional LANs to introduce a new level of concurrency in the communication medium, which becomes an active player in the act of communication (for instance it carries free and busy slots, tokens, etc.).The basic aspect for the broadband ATM LANs and B-ISDN is the need to "fill-in" the bandwidth potential of links having capacities in excess of 1Gb/s, and this can be done only by increasing concurrency. From an architectural point of view, a possible solution is to integrate the network in the memory system of a Global Computer, since servicing the network through the I/O mechanisms of computers creates a performance bottleneck and the memory system seems a more natural vehicle for communication. Another solution is pipe-lining and increasing process parallelism at the edges of the network. These approaches introduce a high degree of non-determinism, which make any form of advance resource reservation and end-to-end controls extremely difficult to apply. Since the time scales of non-deterministic concurrent access to the network are a lot smaller than propagation delays, the resource management procedures must also be able to accommodate imprecision of state. The chosen protocols must not rely on global information, must not require advance reservation, and should allow the receiver to "cache" the future unknown state of the sender in order to accommodate imprecision of state.

Traditionally, there are two kinds of guarantees which are addressed by the resource management procedures: guarantees with respect to *delays* and guarantees with respect to *losses*. Queueing delays are becoming an increasingly smaller proportion of the end-to-end delays, which are latency limited. The *better loss protection* guarantees become more significant, an interesting property of the existing LAN environment. Just like in LANs, a *"best-effort"* service with respect to delays is envisaged. The question is whether LAN-like characteristics could be extended to networks where, in contrast to the LAN environment, not only *the topologies are no longer linear* (like in the current ring or bus based LANs), but also:
A. the time scales of concurrent access to the network resources are much shorter than the propagation delays, making global knowledge obsolete
B. the requirements for multimedia traffic complicate the control of the resource

sharing process
C. the switching technology is cell-based ATM space division switching

A desirable feature from the user point of view is also a very *simple specification* of the connection descriptor, *at best no specification at all*. From the network point of view, controls using global knowledge are less attractive than those using *local knowledge*, which is not subject to latency limitations. A fundamental requirement is to provide means to offer *"fill-in" traffic*, in order to exploit the bandwidth potential of the links. Three methods to achieve these trade-offs are listed here, due to space limitation the reader will find the detailed operation in the references.

Best-effort with Backpressure (BEB)

This scheme [RACE_94] performs no reservation. It trades off delay guarantees for zero loss using a hop-by-hop START/STOP backpressure signal. The STOP signal is generated by the downstream node when the buffer content associated with the controlled VC(s) has reached a "high water mark" and the traffic from the upstream node is throttled until the buffer content has decreased to a "low water mark" (see Figure 1).

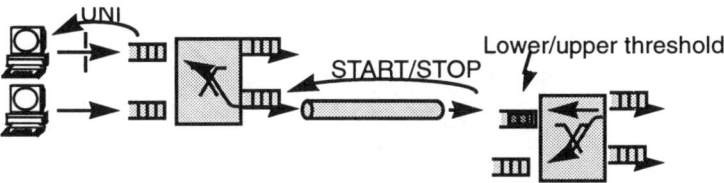

FIGURE 1 Flow control using hop-by-hop backpressure

Fast Reservation Protocols (FRP)

There are two kinds of FRP ([Boy_90]). The one of interest in the context of this paper is the FRP with delayed transmission (FRP-DT) which gives up any guarantee with respect to access delays, but enforces in exchange a zero loss condition: the burst is only admitted on the path when there are enough available resources. The access delay is at least one round-trip delay if the reservation succeeds at the first attempt. The protocol is "rate-based" in the sense that it tries to reserve a certain rate on the link

Hop-by-Hop Flow Control Virtual Channel (FCVC)

This scheme ([KC_93]) performs no reservation. The amount of credit an upstream node can send on a VC to a downstream node is being advertised using a "credit cell" and is computed based on the buffer occupancy at the downstream node. The buffer allocated to a VC is divided in three zones N_1, N_2, N_3 (see Figure 6) and it is the occupancy of the N_2 and N_3 zones (which in turn depends on the traffic intensity on the VC) that is taken into account when deciding the amount of credit. After forwarding N_2 cells from the controlled VC, the downstream node will send a credit cell to the upstream node containing the new credit, which reflects the number of free slots in the N_2+N_3 zone. The method can control traffic on a per VC basis.

5 Quantitative Analysis of Loss-less Statistical Multiplexing Using Discrete Event Simulation

The simulation model has been developed using the discrete-event simulation language CSIM [Sch_91]. CSIM is a process-oriented, general-purpose simulation language written with C language functions. The environment simulated is a quasi parallel execution environment. The basic unit of execution is a process, which can initiate sub-processes. Processes can cause an event to occur, wait for an event to happen, cause and wait for simulated time to pass, request and release facilities and storages. The syntax of the language is normal C (or C++) extended with procedures in a runtime library.

5.1 The Simulation Models

The main concern in order to design the event generator is to be able to map measurable source parameters to mathematically tractable model parameters.

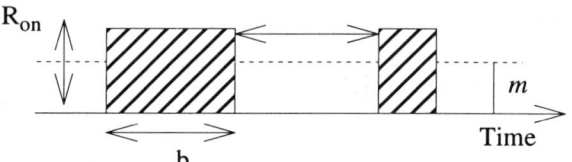

FIGURE 2 Measurable bursty source parameters

It is generally agreed that the peak rate (R_{peak}) of a source is an indispensable parameter and so also one required by the ITU-TS standards on B-ISDN [ITU_92/371] to be declared at call set-up time. A number of proposals include the mean burst period b and either the average bit rate m or the utilization factor ρ as further traffic descriptors. The utilization factor is by definition: $\rho = m / R_{peak}$ and together with R_{peak} yields the mean m and the variance σ^2 of the bit rate, and the burstiness R_{peak}/m. The mean of the burst period b is very useful in describing how bursts are generated by the source and helps discriminate between calls which though having the same peak and mean bit rate, display a different behaviour. For the particular case of exponentially distributed burst and idle periods, it has been demonstrated [GAN_91] that the vector (R_{peak}, ρ, b) completely identifies the traffic statistics of a bursty connection, as depicted in Figure 2. Whatever the approach, the first step needed to derive a solution for the equivalent bandwidth problem is to model the multiple types of sources which are integrated in the broadband network environment. Several models have been proposed in the literature for the analysis of bursty traffic. The models are supposed to combine reasonable accuracy with tractable computation complexity. Though other, more general source models have been proposed, the above requirements lead to the choice of a two-state continuous time Markov chain as in Figure 3. The choice is quite obvious because of two reasons.

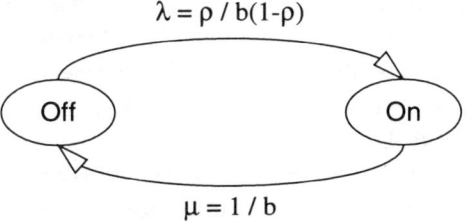

FIGURE 3 A two-state Markov chain source

On one hand, it fits our intuition for a number of widely used traffic sources which have an "on" state (for instance a talk spurt or a burst) and an "off" state (silence or idle). On the other hand, it can be easily expressed in terms of the source parameters, the birth and death rates of a source i are given by:

$$\mu_i = \frac{1}{b_i}, \lambda_i = \frac{\rho_i}{b_i(1-\rho_i)}$$

By varying the parameters R_{on}, T_{on}, T_{off}, different traffic patterns can be generated, as shown in Figure 5. The multiplexer is modelled as one server with to input queues, served in Round-Robin fashion, as shown in Figure 4.

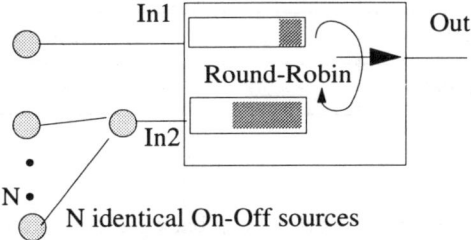

FIGURE 4 The multiplexer

The simulation scenario has to be able to make a comparison between the protocols possible. The main difference is that FRP-DT applies directly to the burst level, whereas BEB and FCVC act on cells. The version of FRP analysed here negotiates service rates with the multiplexer, and these are only indirectly determined by the buffering strategies in the multiplexer. BEB and FCVC on the other hand heavily depend on the size of the $N_1N_2N_3$ zones, even if the buffer occupancy is also influenced by the service rate. Protocol overhead is also an important consideration for the comparison. Since by FRP - DT once a burst is accepted, it will get sent without any protocol overhead, we expect that the overhead will be lower than by BEB. On the other hand, the price for that is that the access delays for FRP are longer for comparable loads. BEB pays the shorter access delays with more bandwidth waisted for the protocol operation. FCVC has an overhead which is uniquely determined by the size of the N_2 zone. So we decided to choose two scenarios. The first one tries to optimize the product (buffer size x overhead) for BEB, by providing in N_2 a number of cells equal to the *(average rate / R_{on})* ratio. By doing so, N_2 will be able to accommodate the mean number of elementary sources simultaneously sending one cell to the buffer. With this choice, simulation has shown that the overhead of BEB is of the order of 2%. The second scenario assumes a buffer of N_2 =50 is used for BEB. This allows to obtain for BEB a bandwidth overhead comparable with FRP-DT on one hand, and a buffer size comparable with FCVC on the other hand. The scenarios are summarized in Table 1.

Table 1: Simulation scenarios

Scenario	Round_trip (cells)	N2
BEB-A	20	m /Ron
BEB-B	20	5*m /Ron
BEB-C	40	m /Ron
BEB-D	40	5*m /Ron
FRP-A	20	N.A.
FRP-C	40	N.A.
FCVC	20	50

5.2   Protocol Behaviour

The analysis focused on the best effort with back pressure protocol. The results for the other protocols have been obtained for comparison purposes only, and do not cover all the scenarios. The simulations focused on the behaviour of the protocols when the offered load shape varies from smooth to bursty, especially the way they insure the "fill-in" behaviour identified in Section 4. In order to substantiate this behaviour, two kinds of conditions were analysed: the case were to average offered load is under the multiplexer capacity, and the case where the average offered load is up to 4 times output link capacity. In the latter case we speak of *sustained overload*. In both cases, the instantaneous peak offered load will depend on the burstiness, and may be as high as 40 times output link capacity. The following parameters were chosen to characterize the two protocols:

-mean protocol induced access delay at the source, i.e. the total amount of time a source which is in the On phase must wait due to Reservation_Denied, or STOP messages, or because of running out of credit respectively (it does not include queueing delays in the multiplexer).

- multiplexer utilization and throughput
- mean queue length in the multiplexer (from which the queueing delays can be derived)
- overhead of the protocol messages as a percentage of the net carried load

Two sources have been multiplexed on the two multiplexer inputs: on In1, one basic ON-OFF background source with exponentially distributed burst and silence periods is sending bursts of mean length 250 kilobytes with a constant R_{on} =10 Mb/s (see Figure 2). By varying the burstiness $(T_{on}+T_{off})/T_{on}$, different mean rates m can be obtained. On the In2 there is a stepwise variable bitrate source as in Figure 5, obtained by superposing of a number of the basic source described above: the rate increments are of R_{on} =10 Mb/s, their duration is exponentially distributed with mean T_{on} = 0.2 s.

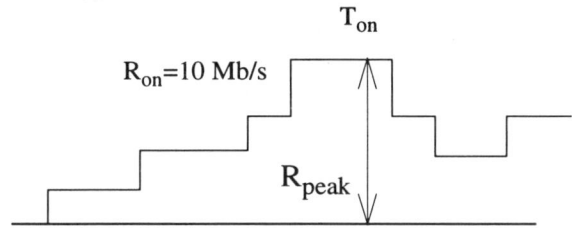

FIGURE 5 Stepwise variable bit rate source

The output link of the multiplexer serves the two inputs in Round Robin fashion.

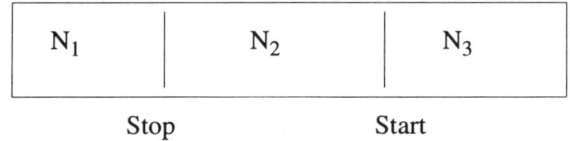

FIGURE 6 Best Effort with Backpressure (BEB) buffer dimensioning

For BEB, the thresholds in the input buffers are dimensioned as following (se Figure 6)
(a) in order to avoid starvation, N_3 = Roundtrip_Delay * Output_Link_Capacity
(b) in order to avoid overflow, N_1=Roundtrip_Delay * R_{peak}
For FRP-DT the buffer was assumed infinite. Roundtrip_Delay was 20 cells. The following scenarios were analysed. In scenario A, the choice was N_2= m/R_{on} [cells], which is a good compromise between a small buffer size and the number of START/STOP signals. Roundtrip_Delay was 20 cells. In scenario B, $N_2' = 5*N_2$, by keeping the Roundtrip_Delay 20 cells. In scenario C, the Roundtrip_Delay was doubled (40 cells) and N_2= m/R_{on}. Finally, in scenario D, both the Roundtrip_Delay and the buffer size had the larger values. In all scenarios, the burstiness was varied from relatively smooth 2 to bursty 50. The scenarios included two overload situations, where the mean offered load is 2.7 and 4 times output link capacity respectively. In all scenarios, the flow of traffic is controlled according to the high and low thresholds for BEB, and to

the rate request and acknowledge cells for FRP-DT. For credit-based flow control, the N_1 and N_3 areas have a similar choice as above. For N_2, the choice is trying to realize a compromise between the buffer size and the resulting protocol overhead. For In1, N_2 = 10 cells (and the overhead 10%), whereas for the link with a much larger bandwidth In2, N_2 = 50 cells (the resulting overhead 2%).

The simulation results show that, even in overload situations the mean access delays are very reasonable and the protocol overhead doesn't go out of control. The individual background source will always get a chance to transmit and even in overload situations the access delay at the input with the background source In1 is zero. The input In2 offering the larger load to the multiplexer shows that the concept of "best-effort" effectively fills-in the bandwidth left free from In1. The utilization of the output link of the multiplexer is 1 for all scenarios summarized in Table 1.

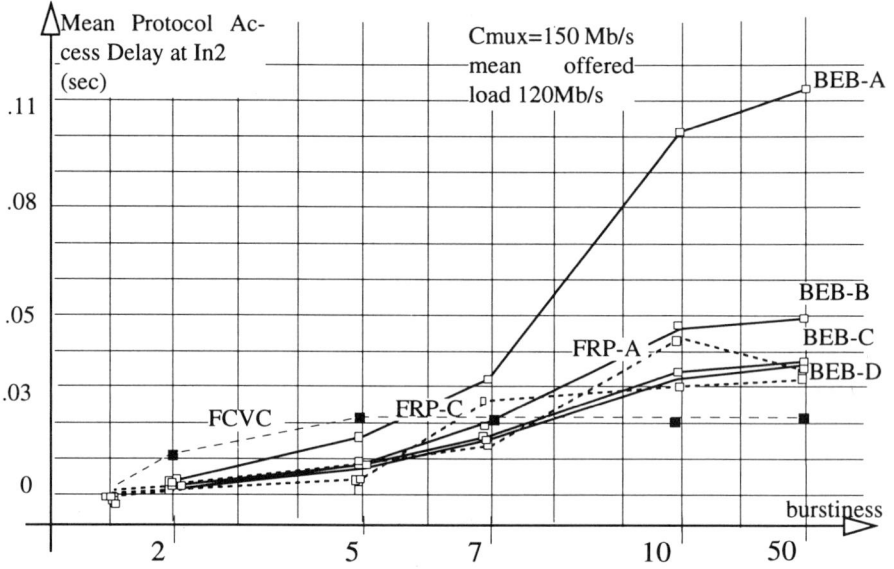

FIGURE 7 Mean protocol access delays, 100000000 events

The comparisons of mean protocol induced delays in Figure 7 show that BEB and FCVC perform generally better then FRP. The behaviour of all protocols is getting worse with the increased burstiness. If the buffer for BEB is just as large to accommodate the mean number of cells generated at In2, then the access delays will increase quite strongly with increased burstiness. If larger buffers comparable with the ones used for FCVC are provided, then the delays are also comparable, but with a smaller bandwidth overhead for BEB. The price for that will be increased waiting times in the multiplexer queue, as shown in Figure 9. In the case of FRP-DT, the individual source is negotiating with the multiplexer rate at each step R_{on} or 0. This "all or nothing" strategy leads to increased mean access delays. This effect is even more dramatic when the offered load is very large, as shown in Figure 8. Average access delays for BEB show a saturation at average load of 600 Mb/s, where they reach approximately 1s. The price for that is a larger bandwidth overhead than FRP-DT. FCVC has a very good performance, due to a better coordination between receiver and sender (the credit update), with the cost of more bandwidth overhead.

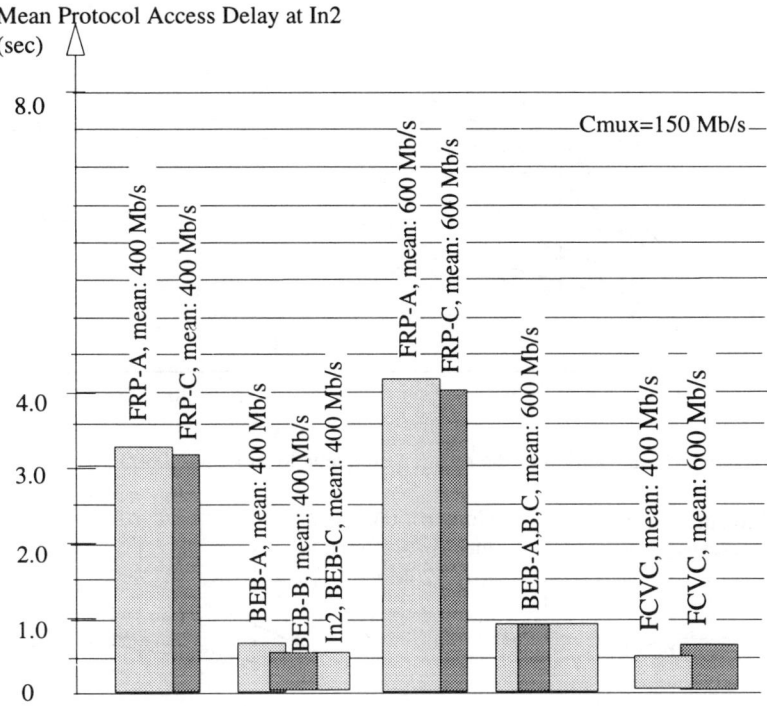

FIGURE 8 Mean access delays under sustained overload, 100000000 events

As expected, the mean queue length increases when the arrival process evolves from "smooth" (burstiness 2) to "unsmooth" (burstiness 50). At high burstiness values the utilization of the source is reduced, and the mean input queue length will no longer increase, as indicated in Figure 9. The overhead due to the protocol messages is remarkably low for FRP-DT, as shown in Figure 10. Under underload conditions it is at about 0.1% and even under overload doesn't get over 6%, as shown in Figure 11. This is the main advantage of the "all or nothing" strategy, since once a burst is accepted, it will get through without interruption. BEB performs also very well by low burstiness but, as illustrated in Figure 10, with higher load and burstiness the overhead increases to about 10%, when the buffer is small. Increasing the buffer size will bring the overhead at about 2% for high burstiness, thus the same value as for FCVC and will outperform FCVC at low burstiness. Finally, the Figure 12 shows very clearly the influence of the buffer size on the multiplexer throughput. A large buffer will always "fill-in" 100% the bandwidth of the output link.

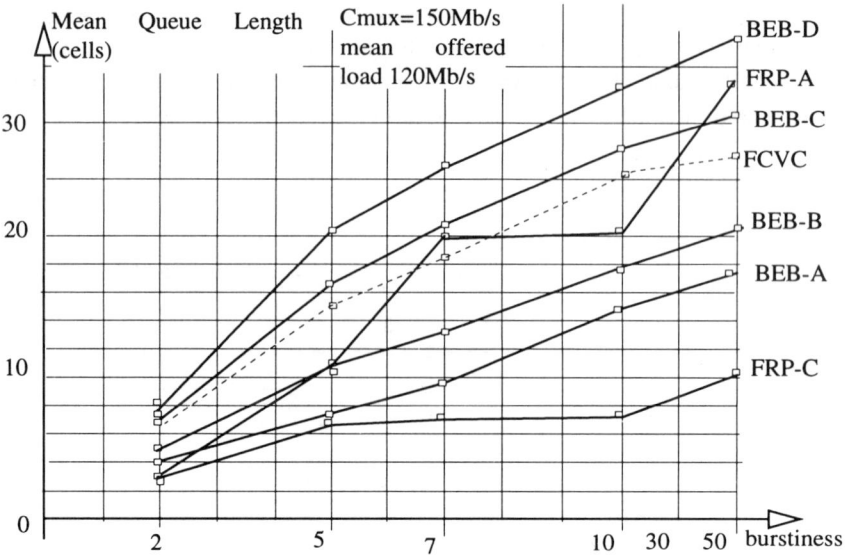

FIGURE 9 .Evolution of the mean input queue length, 100000000 events

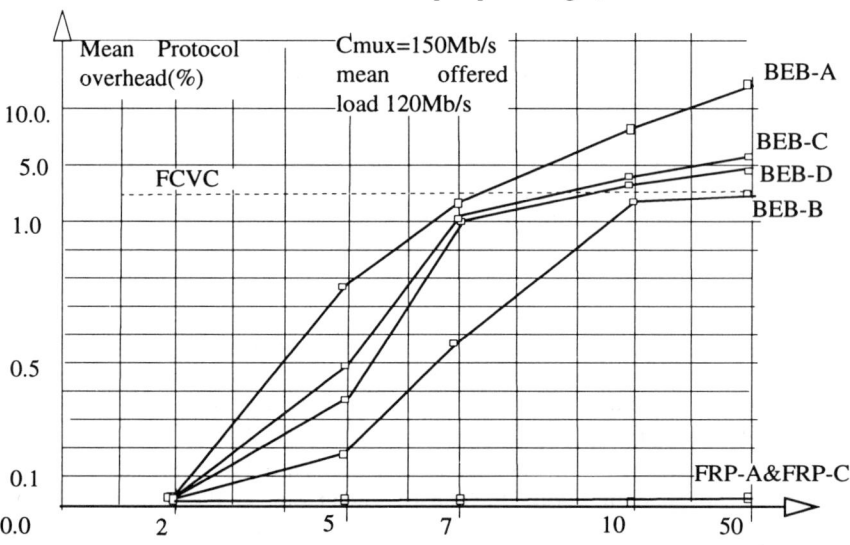

FIGURE 10 Overhead due to protocol messages as percentage of net load, 100000000 events

Trace driven simulation for very bursty traffic

Traffic generated by data applications is extremely bursty. The traffic traces reported in the remarkable papers by Leland [LTWW_93], [FL_91] give an indication on this very important characteristic. In order to asses the suitability of best effort strategies for data traffic, these traces have been used in the following simulations.

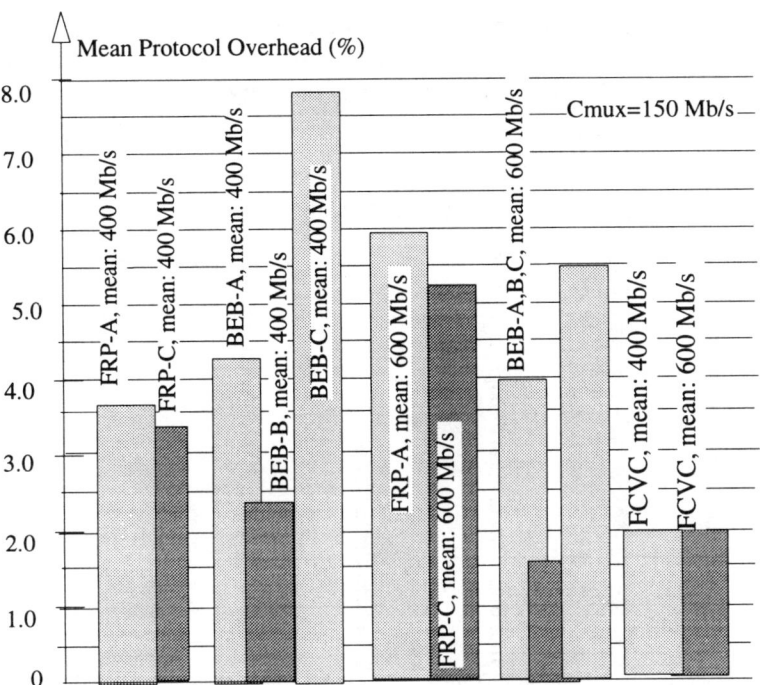

FIGURE 11 Overhead due to protocol messages under sustained overload, 10000000 events

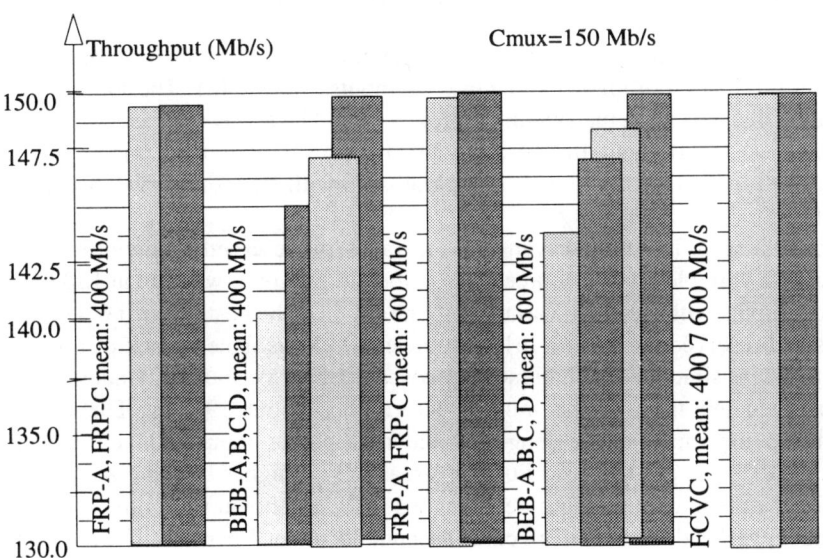

FIGURE 12 Multiplexer throughput under sustained overload, 100000000 events

The particular trace used in the following simulations includes 1 million Ethernet packet arrivals and their respective time stamps. As discussed in [FL_91], the traces show a "fractal like" behavior, i.e. they are bursty on every time scale.

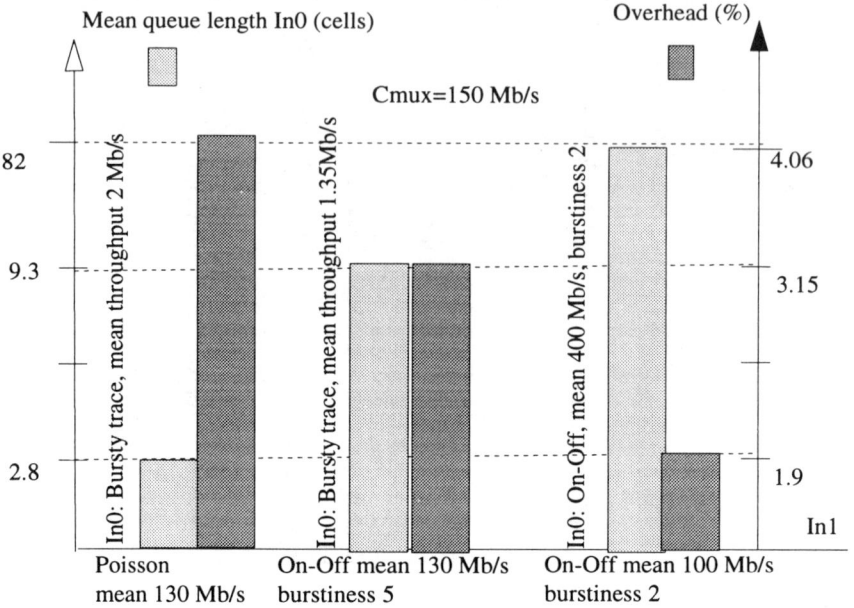

FIGURE 13 Mean queue length and protocol overhead after 1000000 trace cells

In order to quantify the effects of the burstiness, simulations have been performed using a 150 Mb/s multiplexer servicing two inputs: In0 and In1. The sources were at 10 cell time distance from the switch (approx. 9 km). The following cases were analysed:
1. The best effort input In0 driven with the above mentioned traces, the reserved bandwidth input In1 driven by a stream of cells with Poisson arrivals with mean rate 130 Mb/s.
2. The best effort input In0 driven with the traces, the reserved bandwidth In1 driven by a stepwise variable bit rate source as described in Figure 5, with mean 130 Mb/s.
3. Both inputs with stepwise variable bit rate sources, the best effort In0 with mean 400 Mb/s and reserved bandwidth In1 with mean 100 Mb/s. Due to lack of trace data, the overload was simulated with bursty stepwise variable rate sources. Under overload, the mean queueing time in the multiplexer reaches about 191 μs. The backpressure maintains the mean queue length at the multiplexer input In0 at 2.8 cells when multiplexed with the reserved Poisson stream. By a higher burstiness of the reserved traffic, the mean queue length becomes 9.3 cells, as shown in Figure 13. Finally, the protocol overhead remains in all cases acceptable, under 5 percent. The important point here is again that the extreme burstiness of the trace traffic makes it behave at even very low loads comparable with what can be expected from a smother traffic under overload, as can be seen in Figure 13

6 Conclusions

Comparing protocols which provide a similar functionality with different approaches and resources was a challenging task. The decision for the one or the other solution will depend on the weight given to each of the factors analyzed here (access delays, bandwidth overhead, buffer size, link utilization, throughput). The simulations show that if providing large buffers is not considered "expensive", then backpressure with buffer thresholds is a very effective and simple way of resource management. Discrete event simulation is a powerful tool for quantifying the trade-offs associated with these choices, and CSIM was a great help in prototyping a simulation environment.

References

ATM_93   The ATM Forum: "ATM User-Network Interface Specification", Version 3.0, September 1993

Boy_90   P. Boyer, "A congestion control for ATM", presented at the *7th ITC* Specialist Seminar, Morristown, NJ, Oct. 1990, session 4

COST_92 COST 224, "Performance Evaluation and Design of Multiservice Networks", J.W. Roberts, ed., Comm. of European Communities, Brussels (1992)

DMM_91   R. Dighe, C. J. May and G. Ramamurthy, "Congestion avoidance strategies in broadband packet networks", *Proc. of IEEE INFOCOM'91*, paper 4A.1.1, pp. 295-303, 1991

FL_91   H.J. Fowler and W.E. Leland, "Local area network traffic characteristics with implications for broadband network congestion management", in *IEEE JSAC*, vol. 9, no. 7, pp. 1139-1149, Sept. 1991

GAN_91   R. Guerin, H. Ahmadi, and N. Naghshineh, "Equivalent capacity and its application to bandwidth allocation in high-speed networks", in *IEEE JSAC*, vol. 9, no. 7, pp. 968-981, Sept. 1991

HGMY_91 J. Y. Hui, M. B. Gursoy and N. Moayeri, R. Yates, "A layered broadband switching architecture with physical or virtual path configurations", in *IEEE JSAC*, vol. 9, no. 9, pp. 1416-1426, Sept. 1991

Hui_88   J. Y. Hui, "Resource allocation for broadband networks", in *IEEE JSAC*, vol. 6, no. 9, pp. 1598-1608, Dec.1988

ITU_92/371 ITU-TSS "Traffic control and congestion control in B-ISDN", R*ecommendation I. 371*, SG XVIII Temporary Document 1992

KC_93   H.T. Kung and A. Chapman, "The FCVC (flow control virtual channel) proposal for ATM networks", ATM Forum, July 1993

Kle_75   L. Kleinrock, "Queueing Systems", Vol.I: Theory, John Wiley & Sons, New York, 1975

Kel_91   F.P. Kelly, Effective bandwidths at multi-class queues, in *Queueing Systems* vol. 9 pp. 5-16, 1991

LTWW_93 W.E. Leland, M.S. Taqqu, W. Willingger and D.V. Wilson, "On the self similar nature of Ethernet traffic", in Proceedings of ACM SIGCOMM'93, published in Computer Communication Review, vol. 23, no. 4, October 1993

RACE_94 RACE 2068, Business CPN Architecture Functional Specification", Deliverable, January, 1994

Sch_91   H. Schwetman, CSIM Reference Manual, Revision 15, Microelectronics and Computer Technology Corporation, Austin, Texas, 1991